Unraveling Angular 2

The Ultimate Beginners Guide with over 130 Complete Samples

By István Novák

ISBN: 1539061477

ISBN 13: 978-1539061472

To Henriett, Eszter, and Reka, who supported me in writing this book with their love and appreciation.

Table of Contents

About the Author

Istvan Novak is an associate and the chief technology consultant of SoftwArt, a small Hungarian IT consulting company. He works as a software architect and community evangelist. In the last 25 years, he participated in more than 50 enterprise software development projects. In 2002, he co-authored the first Hungarian book about .NET development. In 2007, he was awarded with the Microsoft Most Valuable Professional (MVP) title, and in 2011 he became a Microsoft Regional Director.

As the main author, he contributed in writing the Visual Studio 2010 and .NET 4 Six-In-One book (Wiley, 2010), and in Beginning Windows 8 Application Development (Wiley, 2012). He was the main author of Beginning Visual Studio LightSwitch Development book (Wiley, 2011). In 2014, Istvan started self-publishing his Unraveling series on Amazon.

István holds master's degree from the Technical University of Budapest, Hungary, and also has a doctoral degree in software technology. He lives in Dunakeszi, Hungary, with his wife and two teenage daughters. He is a passionate scuba diver. You may have a good chance of meeting him underwater at the Red Sea in any season of the year.

Introduction

I started my career as a programmer and have been working on the server side of the world for a long time. As web and mobile apps became popular, I moved with my team from desktop-based UI to systems with HTML and JavaScript frontends. It took a long time to learn how to warp the previously used application architectures from fat clients to support the web and mobile interfaces.

One day about three years ago, my team started to work with Angular. In a few weeks, we became as productive with HTML UI, as we ever wanted to be. I was elated and create the "Unraveling Angular 1.3" book so that developers could learn Angular in a natural and relaxed way.

From my point of view, the book was successful. I updated it several times in the last two years. According to readers' reviews, its value was the practical and straightforward style it followed.

In this book, I share my experiences with Angular 2, the second major version the Angular framework.

Who This Book Is For

I wrote this book for everyone who wants to get acquainted with Angular 2 in a very practical way, right from the beginnings. Because you are reading this book, I suppose you have experiences with creating web pages, and some basic knowledge of HTML, CSS, and JavaScript. Do not feel intimidated, if you do not have profound experiences with the nitty-gritty details of these technologies. Provided, you know the basics of creating a very simple web page, and you can host it on a webserver running on your local machine (even using Windows, Mac or Linux), you are ready to go on with the book.

I do not assume that you have ever used Angular 1.x. This book teaches you all essential concepts from scratch. If you already have previous experience with Angular 1.x, it will help you progress a bit faster.

The Angular team develops the new version with TypeScript. Accordingly, in this book I present all code samples in this excellent and promising programming language. Do not get annoyed that you have to learn TypeScript before reading this book! Having a bit more than basic experience with JavaScript is enough to understand the code samples. You will learn the gist of TypeScript— I help you with brief explanations.

Downloading the Source Code

This book provides many examples. You can download the source code as a single zipped file from this link: **http://tinyurl.com/unravelingAngular2**. The link contains separate folders for the different Angular 2 versions. I wrote this edition of the book with the **2.3.0** version.

Similarly to my previous book, I plan to update the chapters and the samples as new releases of the Angular framework are available.

Contact Me

If you have any question, feedback, a recommendation regarding this book, do not hesitate to contact me at **dotneteer@hotmail.com**.

Part I: An Overview of Fundamental Angular 2 Concepts

Chapter 1: A Short Tour of Angular 2

WHAT YOU WILL LEARN IN THIS CHAPTER

Preparing a simple study environment that you can use with this book

Understanding that Angular uses a different approach than the traditional imperative programming

Creating a simple Angular component

Getting acquainted with a couple of fundamental Angular concepts

Angular 2 is an excellent framework; it adds much value—mostly simplicity and productivity—to your everyday tasks with web pages. There are many ways to whet your appetite and let you understand why Angular 2 should get a cushioned cell in your toolbox. I believe, the best way is to show it in action.

Angular 2 is a brand new reincarnation of Angular JS, known as Angular 1.x. The two technologies have a common root, but Angular 2 is built on a more state-of-the-art fashion. In this book, I focus on Angular 2 and refer to Angular JS 1.x only occasionally. Henceforward, I will mention the new framework only as *Angular*.

Get ready, in this chapter, you are going to build a few simple web pages with Angular. I hope, by the time you complete it, you will get flabbergasted—"Is it so easy?"

Preparing Your Study Environment

I want to provide you an easy-to-use study environment with live coding so that you can quickly try the exercises of the book and play with them. You need four software components to follow the samples: a code editor, a web server, a browser, and a few tools.

To edit the code, you can use an integrated development environment such as Webstorm, Visual Studio, Eclipse, or even utilize a simple code editor such as Sublime Text, TextMate, Atom, Notepad++, Visual Studio Code, Brackets, or whatever you prefer. When writing this book, I used Visual Studio Code (https://code.visualstudio.com/), which is a free tool available for Mac, Linux, and Windows.

> ***HINT****: Use an editor that supports syntax highlighting for HTML, CSS, and JavaScript, because this feature makes your life easier.*

To play with the samples, you need a web browser. You can use your preferred browser—Chrome, Firefox, Safari, Internet Explorer, Microsoft Edge, Opera, or other—depending on their availability on your computer. I tested all examples on Chrome, Microsoft Edge, and Internet Explorer 11.

The trinity of tools would not be complete without a web server. Although you can choose from a broad range of them, I recommend you using Node.js with **lite-server**, which is a little development server with live reload capability—and it will help you a lot while learning Angular. I built the toolset utilized in this book on Node.js.

The Samples Working Folder

While you are learning Angular, you will examine code samples. To save you time, I established a sound structure. The root is a folder—I will just refer to it as **Samples**—, which contains subfolders for each chapter. The names of chapter folders follow the convention **ChapterNN**, where **NN** is a two-digit number. Within the chapter folders, you can find subfolders for exercises. For example, the third exercise of Chapter 4 is in the **Samples/Chapter04/Exercise-04-03** folder. Each folder contains every source files—HTML, CSS, and JavaScript (TypeScript) files—you need to run it.

The root folder contains several configuration files—most importantly, **package.json**—that help you install every component you need while working with the samples.

> **NOTE**: *In* Appendix B, *I detail the steps to create these configuration files.*

I recommend you to download the sample code of this book and extract the content of the downloaded **.zip** file into your preferred working folder.

If you favor typing in the code of the samples, create your samples folder. Follow the naming conventions of folders—**ChapterNN/ExerciseNN-MM**—so that the tools for live coding keep working.

Installing Node.js and Configuring lite-server

As I mentioned, this book uses Node.js to build and host the exercise pages. You can download the appropriate binaries for your platform from **https://nodejs.org/en/download**. Run the setup kit, and complete the installation. It takes only a minutes or two.

Installing Tools

The team at Google has been developing Angular 2 with TypeScript (http://typescriptlang.org). Most examples in this book use the TypeScript language, which is a typed superset of JavaScript, and compiles to plain JavaScript.

With the right toolset, you can focus on writing your Angular apps and instantly use them in the study environment. Please follow these steps to set up the tools:

1. Start the command prompt and select **Samples** as your working folder.

2. Install the following tools with Node.js Package Manager (**npm**):

```
npm install typescript -g
npm install typings -g
npm install lite-server -g
npm install concurrently -g
npm install gulp -g
```

3. Install local packages:

```
npm install
```

4. Take care that the core files are updated in each exercise folder:

```
gulp boilerplate
```

In step 2, you installed several tools:

#1: **typescript.** Most examples in this book use the TypeScript language. This tool installs all components you need to work with TypeScript.

#2: **lite-server** is a lightweight web server with live reload capability. It keeps your browser synchronized as you modify the source files.

#3: **concurrently** is a package that allows running multiple commands concurrently. I use this utility in the book to run **lite-server** and the TypeScript compiler simultaneously.

I implemented a few housekeeping tasks with **gulp** that is a simple streaming build system.

In step 3, you installed the files and packages you need to run exercises. You can find these files in the **node_modules** and **typings** folders within **Samples**.

In step 4, you ran the **gulp** utility with the **boilerplate** command. This command takes care that core files are refreshed in each exercise folder.

Testing the Installation

In the **Exercise-01-00** folder, you find a simple web page that can be used to test the installation. Follow these steps:

1. Start the command prompt and select **Samples/Chapter01/Exercise-01-00** as your working folder.

2. Type the **npm start** command.

If you managed to install all tools and components successfully, you see the welcome message on the page (Figure 1-1).

Figure 1-1: The message indicates successful installation

3. Close the browser and terminate the running command line (Ctrl+C).

> ***NOTE****: This simple page does not use Angular 2 yet. Nonetheless, is uses a short TypeScript code snippet to display the message. With the steps above you can test the successful installation and configuration of the* **concurrently***,* **lite-server***,* **typescript***, and* **typings** *components.*

Housekeeping Utilities

As new versions of Angular 2 and its dependent components are released continuously, you may need to change the content of configuration files, such as **package.json**, and the others. Because this book has more than a hundred exercises, it would be painful and laborious to traverse through all exercise folders and carry out the modifications manually. You can do it easier and quicker:

1. Modify these files directly in the **Samples** folder.

2. Run the **gulp boilerplate** command line from within the **Samples** folder. This command will update the content of each exercise folder with the configuration files you modified in **Samples**.

You can use the **gulp copy** task to create a copy of an exercise, but you only need to invoke this task if you decide to create the exercises manually. For example, if you want to build **Exercise 01-02** as a copy of **Exercise 01-01**, use this command:

```
gulp copy --ex 01-01
```

If the **Exercise-01-02** folder does not exist, this task creates it and copies all files from **Exercise-01-01** into it.

> ***NOTE****: You can find these tasks in* **Samples/gulpfile.js***.*

Creating an HTML Page with Scuba Dive Log Records

Now that you a have a study environment, it is time to make the first steps to get acquainted with Angular. Before you create your first web page that leverages Angular, let's start with plain HTML.

In this book, you will create prototype HTML pages for a fictitious website, *Younderwater*, which is a hub for scuba divers from all around the world. Creating such a website can be challenging. This book will focus on how you can leverage Angular to create a few pages and components of this site.

Each qualified scuba diver must keep a log about his or her dives. A log record may contain dozens of fields, but only a few of them are required in an official logbook. To create a simple HTML form that displays a couple of dive log records, follow these steps:

1. Create a new folder, **Exercise-01-01**, under your working folder (remember, your working folder is **Samples/Chapter01**).

2. Copy **package.json** and **bs-config.json** from **Samples** to **Exercise-01-01**.

3. In this new folder, create an **index.html** file with the code in Listing 1-1.

Listing 1-1: index.html (Exercise-01-01)

```
<!DOCTYPE html>
<html>
<head>
  <title>Dive Log (HTML)</title>
  <link href="/node_modules/bootstrap/dist/css/bootstrap.min.css"
    rel="stylesheet" />
</head>
<body>
  <div class="container-fluid">
    <h1>My Latest Dives (HTML)</h1>
    <div class="row">
      <div class="col-sm-4">
        <h3>Abu Gotta Ramada</h3>
        <h4>Hurghada, Egypt</h4>
        <h2>72 feet | 54 min</h2>
      </div>
      <div class="col-sm-4">
        <h3>Ponte Mahoon</h3>
        <h4>Maehbourg, Mauritius</h4>
        <h2>54 feet | 38 min</h2>
      </div>
      <div class="col-sm-4">
        <h3>Molnar Cave</h3>
        <h4>Budapest, Hungary</h4>
        <h2>98 feet | 62 min</h2>
      </div>
    </div>
  </div>
```

```
    </div>
  </body>
</html>
```

This page contains three dive log records baked into the markup. It also uses the Bootstrap stylesheet file.

4. From the exercise folder, start the app:

```
cd Chapter01/Exercise-01-01
npm run lite
```

The **npm run lite** command line starts the app in your default browser, and the page displays the dive log entries, as shown in Figure 1-2.

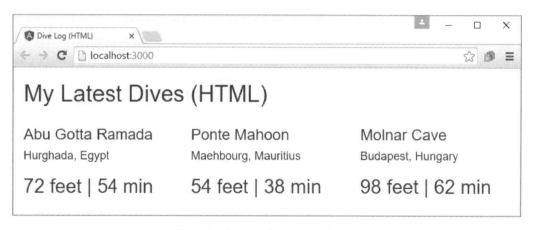

Figure 1-2: The dive log entries in the page

If you have not used Bootstrap before, here is a brief summary about it:

Bootstrap defines a set of great CSS styles that support its flexible layout model—described with the **<div>** tags in Listing 1-1. The outermost **<div>** defines a container that changes its size with the browser window. The **<div>** with the **row** style represents a single row grid (12 units wide), and each nested **<div>** with the **col-sm-4** style marks a single cell (4 units wide). If you want to learn more details on Bootstrap, let me recommend you my *Unraveling Bootstrap 3.3* book.

> **HINT:** *You can terminate **lite-server** with Ctrl+C.*

Using jQuery to Build the Dive Log Page

The page you created is a static HTML page, with the markup repeated for each dive log entry. In the real world, these records come from the backend, and—in most cases—you display them on the client side with JavaScript code.

Let's change the previous example to render the markup dynamically with jQuery. You can find this solution in the **Exercise-01-02** folder. If you want to create the page manually, follow these steps:

1. Use this command line to create a copy of the **Exercise-01-01** folder into **Exercise-01-02**:

```
gulp copy --ex 01-01
```

2. Create a new file, **dives.js**, in the new folder and type the code in Listing 1-2 into it.

Listing 1-2: dives.js (Exercise-01-02)

```
var dives = [
  {
    site: 'Abu Gotta Ramada',
    location: 'Hurghada, Egypt',
    depth: 72,
    time: 54
  },
  {
    site: 'Ponte Mahoon',
    location: 'Maehbourg, Mauritius',
    depth: 54,
    time: 38
  },
  {
    site: 'Molnar Cave',
    location: 'Budapest, Hungary',
    depth: 98,
    time: 62
  }];
```

This file declares the **dives** array and initializes it with the dive log data that the sample will display on the page.

3. Remove the markup from within the **<div>** tag with the **row** style and add the **id** attribute with the **logbook** value to it. Add the **<script>** sections highlighted in Listing 1-3 to the page, along with the other small changes, too.

Listing 1-3: index.html (Exercise-01-02)

```
<!DOCTYPE html>
<html>
<head>
  <title>Dive Log (jQuery)</title>
  <link href="./node_modules/bootstrap/dist/css/bootstrap.min.css"
    rel="stylesheet" />
</head>
<body>
  <div class="container-fluid">
```

25

```
    <h1>My Latest Dives (jQuery)</h1>
    <div class="row" id="logbook">
    </div>
  </div>

  <script src="./node_modules/jquery/dist/jquery.min.js"></script>
  <script src="dives.js"></script>
  <script>
    $(document).ready(function () {
      for (var i = 0; i < dives.length; i++) {
        var dive = dives[i];
        $("#logbook").append(
          "<div class='col-sm-4'>" +
          '<h3>' + dive.site + '</h3>' +
          '<h4>' + dive.location + '</h4>' +
          '<h2>' + dive.depth + ' feet | ' +
          dive.time + ' min</h2>' +
          '</div>');
      }
    });
  </script>
</body>
</html>
```

Run the app from the **Exercise-01-02** folder:

```
npm run lite
```

As shown in Figure 1-3, dive log entries are now built by the script you added in the previous step.

> **HINT:** *If you're browser displays a "Cannot GET /" message, probably you forgot to select the exercise folder* (**Chapter01/Exercise-01-02**) *before invoking* **npm run lite**.

Although this code is simple, I would not say it is easy to maintain. First, it uses imperative approach and dynamically extends the DOM with the help of jQuery. Second, it is not complex, but still not pretty straightforward to guess out what kind of markup will the jQuery transformation result. In the real life, the markup describing a log record may be more complex, and the "let's put it together with jQuery primitives manipulating the DOM" approach is pretty challenging.

Figure 1-3: The page built with jQuery

Moving to Angular

In contrast to the imperative style of jQuery, Angular provides a declarative approach to creating your apps. As of today, you can use three programming languages to write Angular apps: vanilla JavaScript (ES5 or ES2015), TypeScript, and Dart. In this book, I will use TypeScript because it helps leverage you teach the benefits of Angular in a bit more expressive way than JavaScript. Though Dart is an excellent language, I have only a little experience with it, so I will not treat it.

JavaScript evolved a lot in the last five years. Today all the key browsers handle the ES5 (ECMAScript 5) version of JavaScript, and most of them support the newest JavaScript version, ES2015 (ECMAScript 2015).

The significance of ES2015 is that it has great new features, such as classes, modules, promises, iterators, generators, and others, which make it a more productive programming language with all the power its predecessor has.

The Angular team at Google decided to leverage this power. They have been implementing Angular 2 for years with the TypeScript programming language. TypeScript is a typed superset of JavaScript, and it can compile TypeScript programs to JavaScript—including ES5 or ES2015.

Although Angular is called a framework, it continuously shifts to becoming a complete development platform for the web, mobile, and desktop. To allow you start with simple apps and make them more compound and complex, you need to provide a few constructs to run the purest app.

Creating the Dive Log Entry App

In the next exercise, we move the dive log entry app into Angular—with TypeScript.

> **HINT:** *Do not be intimidated if you do not know TypeScript. You will be able to understand the sample. In a few sections later in this chapter, you will get acquainted with the most fundamental concepts of TypeScript. In the subsequent chapters, I will treat any new feature that you need to grab to follow a particular exercise.*

You can find the completed solution in the **Exercise-01-03** folder. Even if you do not want to create the app manually, take a look at these steps:

1. Use this command line to create a copy the **Exercise-01-02** folder into **Exercise-01-03**:

```
gulp copy --ex 01-02
```

2. Copy the **package.json**, **tsconfig.json**, **systemjs.config.js** and **bs-config.json** files there from the **Samples** folder into **Exercise-01-03**.

3. Type the code in Listing 1-4 into **index.html**.

Listing 1-4: index.html (Exercise-01-03)

```html
<!DOCTYPE html>
<html>
<head>
  <title>Dive Log (Angular/TypeScript)</title>
  <link href="./node_modules/bootstrap/dist/css/bootstrap.min.css"
    rel="stylesheet" />
  <script src="/node_modules/core-js/client/shim.min.js"></script>
  <script src="/node_modules/zone.js/dist/zone.js"></script>
  <script src="/node_modules/reflect-metadata/Reflect.js"></script>
  <script src="/node_modules/systemjs/dist/system.src.js"></script>

  <script src="systemjs.config.js"></script>
  <script>
    System.import('app').catch(function(err){ console.error(err); });
  </script>
</head>
<body>
  <divelog>Loading...</divelog>
</body>
</html>
```

4. Create an **app** folder under **Exercise-01-03**; this will contain the application-specific files.

5. Create a new file, **app/dive-log.component.ts** to declare the main class of your app, with the content shown in Listing 1-5.

Listing 1-5: dive-log.component.ts (Exercise-01-03)

```typescript
import {Component} from '@angular/core';

@Component({
  selector: 'divelog',
  templateUrl: 'app/dive-log.template.html'
})
export class DiveLogComponent {
  public dives = [
    {
      site: 'Abu Gotta Ramada',
      location: 'Hurghada, Egypt',
      depth: 72,
      time: 54
    },
    {
      site: 'Ponte Mahoon',
      location: 'Maehbourg, Mauritius',
      depth: 54,
      time: 38
    },
    {
      site: 'Molnar Cave',
      location: 'Budapest, Hungary',
      depth: 98,
      time: 62
    }];
}
```

6. Create a template file (app/dive-log.template.html) with the markup in Listing 1-6.

Listing 1-6: dive-log.template.html (Exercise-01-03)

```html
<div class="container-fluid">
  <h1>My Latest Dives (Angular/TypeScript)</h1>
  <div class="row">
    <div class="col-sm-4"
      *ngFor="let dive of dives">
      <h3>{{dive.site}}</h3>
      <h4>{{dive.location}}</h4>
      <h2>{{dive.depth}} feet | {{dive.time}} min</h2>
    </div>
  </div>
</div>
```

7. Create an application root module (`app/app.module.ts`) with the content in Listing 1-7.

Listing 1-7: app.module.ts (Exercise-01-03)

```
import {NgModule} from '@angular/core';
import {BrowserModule} from '@angular/platform-browser';

import {DiveLogComponent} from './dive-log.component';

@NgModule({
  imports: [BrowserModule],
  declarations: [DiveLogComponent],
  bootstrap: [DiveLogComponent]
})
export class AppModule { }
```

8. Add the **main.ts** file to the **app** folder, as shown in Listing 1-8.

Listing 1-8: main.ts (Exercise-01-03)

```
import {platformBrowserDynamic} from '@angular/platform-browser-dynamic';
import {AppModule} from './app.module';

platformBrowserDynamic().bootstrapModule(AppModule);
```

9. Now, everything is put together, you can run the app from the **Exercise-01-03** folder:

```
npm start
```

Observe, this time, you run the app with the **npm start** command line instead of **npm run lite**. The reason is that **npm start** invokes the TypeScript compiler (**tsc**) in parallel with **lite-server**, and it is required in order the **main.ts** file would be transpiled to JavaScript (**main.js**). The heading text shows that now the Angular/TypeScript version runs (Figure 1-4).

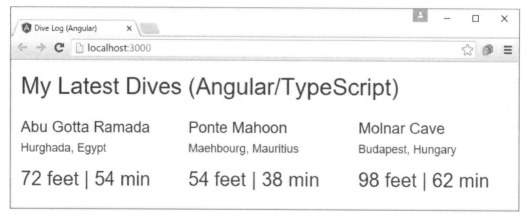

Figure 1-4: The page built with Angular

How This App Works

Let's dive into a few operation details of this app! As the markup of **index.html** shows, we include several script files from the **node_modules** folder to make the bed for Angular:

```
<script src="/node_modules/core-js/client/shim.min.js"></script>
<script src="/node_modules/zone.js/dist/zone.js"></script>
<script src="/node_modules/reflect-metadata/Reflect.js"></script>
<script src="/node_modules/systemjs/dist/system.src.js"></script>
```

This exercise uses Systemjs (https://github.com/systemjs/systemjs), a universal dynamic module loader, which supports ES2015 modules, AMD, and Commonjs as well as global scripts in the browser and Node.js.

The highlighted line shows that Systemjs is loaded. This script configures the loader and prepares it to handle module requests:

```
<script src="systemjs.config.js"></script>
<script>
  System.import('app').catch(function(err){ console.error(err); });
</script>
```

The **systemjs.config.js** file sets up mappings and packages to tell the module loader where to look for certain packages, and how to load them. The **System.config()** call instructs the module loader to import the **app** module. According to the configuration in **systemjs.config.js**, the **app/main.js** file is loaded. Because **tsc** already compiled **main.ts** to **main.js**, **app/main.js** can be loaded and executed.

> **NOTE**: *You will learn more details about Systemjs later in this book, in* Chapter 5, How Systemjs Works

Now, you already know how the app loaded its simple module file (**main.ts**). Let's see, how the TypeScript code moves the gears of Angular!

```
import {platformBrowserDynamic} from '@angular/platform-browser-dynamic';
import {AppModule} from './app.module';

platformBrowserDynamic().bootstrapModule(AppModule);
```

The two **import** statements—with the help of the TypeScript compiler—request the loader to import the modules named in the **from** clauses. The loader uses its configuration—remember, it is set in **systemjs.config.js**—to find out how to load and instantiate the modules. The **platformBrowserDynamic** and **AppModule** exports of the loaded modules become available in **main.ts**.

The **platformBrowserDynamic** object is a function, and with its help, Angular loads **AppModule** and uses the module's metadata to carry out all the chores to bootstrap **AppComponent**.

Angular utilizes the metadata assigned to objects to create and use them. In the code, **AppModule** is decorated with the **NgModule** decorator—this is what **@NgModule(...)** means. **NgModule** receives a configuration object with three properties, **imports**, **declarations**, and **bootstrap**, respectively:

```
@NgModule({
  imports: [BrowserModule],
  declarations: [DiveLogComponent],
  bootstrap: [DiveLogComponent]
})
export class AppModule { }
```

This metadata tells the framework that this app imports a single helper module, **BrowserModule**, which is a core part of the Angular framework that all browser app must import. Besides, the app declares a single component, **DiveLogComponent**, which is the one to be bootstrapped when the page is loaded.

> **TypeScript**: As ES 2015 defines this role, decorators add metadata to classes, properties and object literals at design time with a declarative syntax. At run time, this metadata can be queried, tested, and an object can modify its behavior accordingly. As this exercise proves, TypeScript implements decorators.
>
> Angular utilizes TypeScript classes to define its building blocks, such as modules, components, and others.

In the code, the **DiveLogComponent** class has metadata, too. It is decorated with the **Component** decorator—this is what **@Component(...)** means—that receives a configuration object with a **selector** and a **templateUrl** property:

```
@Component({
  selector: 'divelog',
  templateUrl: 'app/dive-log.template.html'
})
class DiveLogComponent {
  // ...
}
```

So we have learned that **main.ts** drives Angular to load **AppModule** and bootstrap **DiveLogComponent**. This is the moment where the framework examines the metadata of **DiveLogComponent** to decide how to handle it:

"Let's see this **DiveLogComponent** guy! What metadata does it have? Oh, it has a **Component** decorator, so it must be a component. The decorator says that it has a **selector** of "divelog", and an external template in the **app/dive-log.template.html** file.

I see one occurrence of **<divelog>** element in **index.html**, so let's create a component instance for it using the template in that file!"

The body of **index.html** is pretty concise, it contains a simple **<divelog>** element:

```
<body>
  <divelog>Loading...</divelog>
</body>
```

DiveLogComponent defines the behavior of the **<divelog>** tag—as if you had extended HTML with it— so that it displays the three dive log element. Let's examine, how the code does it!

Angular has a central concept, *component*. A component controls a part of the page and defines the logic responsible for that speck of UI. In the code, the **DiveLogComponent** object represents such a component. Its class is decorated with this metadata object:

```
{
  selector: 'divelog',
  templateUrl: './divelog-template.html'
}
```

Here, the value of **selector** tells the framework that an instance of this component should be attached to each **<divelog>** HTML element in the page. The **dive-log.template.html** file declares the template of the UI.

DiveLogComponent initializes its **dives** property with an array of dive log entries:

```
export class DiveLogComponent {
  public dives = [
    {
      site: 'Abu Gotta Ramada',
      location: 'Hurghada, Egypt',
      depth: 72,
      time: 54
    },
    // ...
  ];
}
```

Angular transforms the template of **DiveLogComponent** into a view—a partial DOM fragment—and displays it:

```
<div class="container-fluid">
  <h1>My Latest Dives (Angular ES5)</h1>
  <div class="row">
    <div class="col-sm-4
      *ngFor="let dive of dives">
      <h3>{{dive.site}}</h3>
      <h4>{{dive.location}}</h4>
      <h2>{{dive.depth}} feet | {{dive.time}} min</h2>
    </div>
  </div>
</div>
```

The rendering engine parses this template, and recognizes the ***ngFor** attribute of **<div>** as a *directive*, and the **{{dive.site}}**, **{{dive.location}}**, **{{dive.depth}}** and **{{dive.time}}** expressions as *interpolated strings*. The ***ngFor** directive iterates through the dive log entries stored in the **dives** property of the component, the current iteration is stored in the **dive** local variable. When the engine evaluates the interpolated strings—it observes them by the "**{{**" prefix and "**}}**" suffix—the properties of the dive local variable are converted to their string representation.

At the end of the day, the framework injects the DOM of the view into the **<divelog>** element, as shown in Figure 1-5, and this is how the page displays the dive log entries.

```
▼ <divelog>
   ▼ <div class="container-fluid">
        ::before
        <h1>My Latest Dives (Angular)</h1>
      ▼ <div class="row">
           ::before
           <!--template bindings={}-->
         ▼ <div class="col-sm-4">
              <h3>Abu Gotta Ramada</h3>
              <h4>Hurghada, Egypt</h4>
              <h2>72 feet | 54 min</h2>
           </div>
         ▼ <div class="col-sm-4">
              <h3>Ponte Mahoon</h3>
              <h4>Maehbourg, Mauritius</h4>
              <h2>54 feet | 38 min</h2>
           </div>
         ▼ <div class="col-sm-4">
              <h3>Molnar Cave</h3>
              <h4>Budapest, Hungary</h4>
              <h2>98 feet | 62 min</h2>
           </div>
           ::after
        </div>
        ::after
     </div>
  </divelog>
```

Figure 1-5: The DOM rendered by Angular

Filtering Dive Log Entries with a Pipe

When the logbook contains dozens of dives, having a search field may help a lot to find entries you are looking for. Angular has a concept, *pipe*, which takes data input and transforms it to another output.

In the next exercise, you are going to create a new pipe that takes the list of dive log entries as its input and transforms this list into a filtered one. The app will provide a search field, and displays only those items that have any value matching with the search field. Figure 1-6 shows the app in action.

Figure 1-6: Filtering dive log entries

To update the previous sample, follow these steps:

1. Copy the content of the **Exercise-01-03** folder into **Exercise-01-04**, and work within the new folder:

```
gulp copy --ex 01-03
```

2. Change the **dive-log.template.html** file to add the markup for the search box, and enable filtering by the content of the text typed into the search box, as shown in Listing 1-9:

Listing 1-9: divelog-template.html (Exercise-01-04)

```
<div class="container-fluid">
  <h1>My Latest Dives (Angular/TypeScript)</h1>
  <div class="row">
    <div class="col-sm-4 col-sm-offset-8">
      <input #searchBox class="form-control input-lg"
        placeholder="Search"
        (keyup)="0" />
    </div>
  </div>
  <div class="row">
    <div class="col-sm-4"
      *ngFor="let dive of dives | contentFilter:searchBox.value">
      <h3>{{dive.site}}</h3>
      <h4>{{dive.location}}</h4>
      <h2>{{dive.depth}} feet | {{dive.time}} min</h2>
    </div>
  </div>
</div>
```

3. Add a new file, **content-filter.pipe.ts** to the app folder with the content in Listing 1-10.

Listing 1-10: content-filter.pipe.ts (Exercise-01-04)

```
import {Pipe, PipeTransform} from '@angular/core';

@Pipe({name: 'contentFilter'})
export class ContentFilterPipe implements PipeTransform {
  transform(value: any[], searchFor: string) : any[] {
    if (!searchFor) return value;
    searchFor = searchFor.toLowerCase();
    return value.filter(dive =>
      dive.site.toLowerCase().indexOf(searchFor) >= 0 ||
      dive.location.toLowerCase().indexOf(searchFor) >= 0 ||
      dive.depth.toString().indexOf(searchFor) >= 0 ||
      dive.time.toString().indexOf(searchFor) >= 0);
  }
}
```

4. Modify the root application module, **AppModule**, with the declaration of **ContentFilterPipe** (Listing 1-11).

Listing 1-11: app.module.ts (Exercise-01-04)

```
import {NgModule} from '@angular/core';
import {BrowserModule} from '@angular/platform-browser';

import {DiveLogComponent} from './dive-log.component';
import {ContentFilterPipe} from './content-filter.pipe'

@NgModule({
  imports: [BrowserModule],
  declarations: [
    DiveLogComponent,
    ContentFilterPipe
  ],
  bootstrap: [DiveLogComponent]
})
export class AppModule { }
```

5. Now you can run the app from the **Exercise-01-04** folder:

```
npm start
```

Try to specify a few different searches and see how the dive log items are filtered by the criteria you type in. You can even use digits to filter the time and depth. For example, when you type "5", the page displays two dive log items that have a "5" in time and depth, respectively (Figure 1-7).

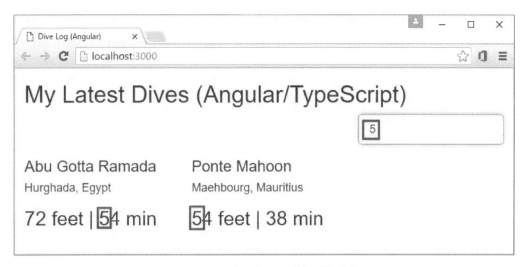

Figure 1-7: Dive log items filtered by "5"

Let's see, how this sample works!

The markup of the search box takes care that the current **value** property of the **<input>** tag can be utilized as the search criterium:

```
<input #searchBox class="form-control input-lg"
   placeholder="Search"
   (keyup)="0" />
```

Here, the **#searchBox** attribute instructs the Angular engine to create a reference to the DOM element that represents **<input>** and store it in the **searchBox** local variable. With this reference, the code can grab the **value** property of **<input>**. The **(keyup)="0"** attribute instructs Angular to execute the "0" expression whenever a "keyup" event occurs on **<input>**. Well, the "0" expression does not do anything, but it lets Angular respond to the event and allows the **value** property of the DOM object referenced by **searchBox** to be updated.

> **NOTE**: *Without the **(keyup)** event binding this sample would not work, because Angular would not detect the change of the search box value.*

The ***ngFor** directive in the template definition contains a pipe expression:

```
*ngFor="let dive of dives | contentFilter:searchBox.value"
```

Pipes are very simple concepts: they can transform objects. The *pipe operator* ("|") signs that we intend to convert the content of **dives** by a pipe named "contentFilter", and that pipe is passed an argument, the value of the "searchBox.value" expression.

This class defines the "contentFilter" pipe:

```
@Pipe({name: 'contentFilter'})
class ContentFilterPipe implements PipeTransform {
  transform(value: any[], searchFor: string) : any[] {
    if (!searchFor) return value;
    searchFor = searchFor.toLowerCase();
    return value.filter(dive =>
      dive.site.toLowerCase().indexOf(searchFor) >= 0 ||
      dive.location.toLowerCase().indexOf(searchFor) >= 0 ||
      dive.depth.toString().indexOf(searchFor) >= 0 ||
      dive.time.toString().indexOf(searchFor) >= 0);
  }
}
```

The **@Pipe()** decorator assigns metadata to the **ContentFilterPipe** class. It declares that this class acts as a pipe, and it should be associated with the "contentFilter" name in Angular expressions. The **ContentFilterPipe** class implements the **PipeTransform** interface, which defines the **transform()** method. When Angular applies a pipe, it passes the value to the pipe (the value to the left to "|") as the first parameter of **transform()**, and then the pipe arguments declared in the pipe expression as subsequent parameters.

> **TypeScript**: *Just as many object-oriented programming languages, TypeScript allows defining and implementing interfaces. As the definition of* **transform()** *shows, you can add optional type notation to parameters. In the code above,* **value** *can be an array of arbitrary objects, while* **searchFor** *is expected to be a string.*

Thus, the "**dives | contentFilter:searchBox.value**" pipe expression passes the **dives** object as the first, and the **searchBox.value** as the second parameter of **transform()**. The pipe filters its input value to the items that contain the search text in any dive log entry field.

In order the **DiveLogComponent** can use the **ContentFilterPipe**, we must tell Angular that such a pipe exists in our module, and this is why we add the pipe's declaration to the root module:

```
@NgModule({
  imports: [BrowserModule],
  declarations: [
    DiveLogComponent,
    ContentFilterPipe
  ],
  bootstrap: [DiveLogComponent]
})
export class AppModule { }
```

Forgetting about this declaration leads to an error message (Figure 1-8).

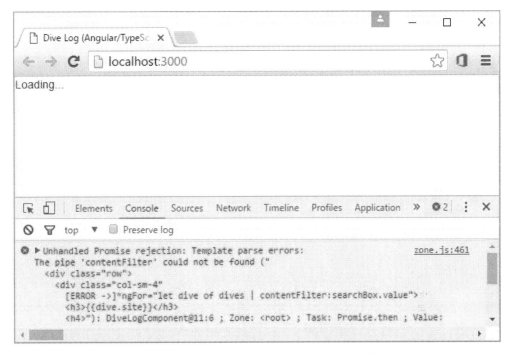

Figure 1-8: Error message about an unknown identifier

Whenever we type a new expression into the search box—every time we press a key—, the Angular change detection mechanism observes that the search box content changes, and so the "`dives | contentFilter:searchBox.value`" expression changes, too. The framework invokes the pipe again and then refreshes the modified content of the page accordingly.

Naming Conventions

When creating the last two samples, you could observe the file naming conventions we used, such as `main.ts`, `dive-log.component.ts`, and `content-filter.pipe.ts`. Not only file names, but class names also followed a particular style, and their names ended with suffixes that represent the role of the class (`Component`, `Pipe`).

These conventions help you follow the structure of the code—especially when it gets bigger and more complex—, file names make it easy to identify what kind of class or other structure they contain.

> **NOTE**: *You can find the latest version of this style guide in this link:*
> https://angular.io/docs/ts/latest/guide/style-guide.html.

Adding Actions to the Component

In the previous exercises, we simply displayed data. Now, it is time to add some action to the tiny scuba dive log app. In this exercise, we are going to add two buttons that mimic entering a new dive log record and clearing the log, respectively. Figure 1-9 shows the app in action.

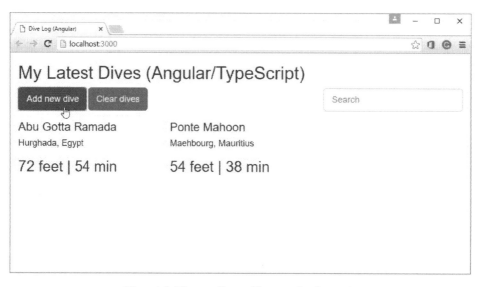

Figure 1-9: The app allows adding new dive log entries

You can find the source code of this sample in the **Exercise-01-05** folder. Now, you are experienced enough to build this app without detailed instructions. To define the new application logic, we should change the **dive-log.component.ts** file, as shown in Listing 1-12.

Listing 1-12: dive-log.component.ts (Exercise-01-05)

```typescript
import {Component} from '@angular/core';

@Component({
  selector: 'divelog',
  templateUrl: 'app/dive-log.template.html'
})
export class DiveLogComponent {
  public dives = [];
  private _index = 0;
  private stockDives = [
    // --- Omitted for the sake of brevity
  ];

  public enableAdd() {
    return this._index < this._stockDives.length;
  }
}
```

```
  public addDive() {
    if (this.enableAdd()) {
      this.dives.push(this._stockDives[this._index++]);
    }
  }

  public clearDives() {
    this.dives = [];
    this._index = 0;
  }
}
```

This app does not allow you to enter new dive log entries; it adds new entries from stock items stored in the **_stockDives** array of the class. The **dives** array stores the entries to be displayed.

The logic is pretty straightforward. The **enableAdd()** function checks if there are any more log entries to add. The **addDive()** and **clearDives()** functions carry out the actions of the Add new site and Clear dives buttons, respectively.

Let's talk about a few details of TypeScript! The variables declared within the class are *class members*:

```
public dives = [];
private _index = 0;
private _stockDives = [
  // ...
]
```

The **dives** member is marked as **public**, so this member is accessible from outside the class by other TypeScript objects—unlike **_index** and **_stockDives**, which are marked as **private**.

Class member function declarations use a straightforward syntax free from unnecessary adornments. Within function bodies, we can refer the members of the same class instance with the **this** keyword:

```
public addDive() {
  if (this.enableAdd()) {
    this.dives.push(this.stockDives[this.index++]);
  }
}
```

To let the app use the actions in the **DiveLogComponent** class, we need to bind them to the component template, as shown in Listing 1-13.

Listing 1-13: dive-log.component.ts (Exercise-01-05)

```
<div class="container-fluid">
  <h1>My Latest Dives (Angular/TypeScript)</h1>
  <div class="row">
    <div class="col-sm-5">
```

```
        <button class="btn btn-primary btn-lg"
          [disabled]="!enableAdd()"
          (click)="addDive()">
            Add new site
        </button>
        <button class="btn btn-danger btn-lg"
          (click)="clearDives()">
            Clear dives
        </button>
      </div>
      <div class="col-sm-4 col-sm-offset-3">
        <input #searchBox class="form-control input-lg"
          placeholder="Search"
          (keyup)="0" />
      </div>
    </div>
    <div class="row">
      <div class="col-sm-4"
        *ngFor="let dive of dives | contentFilter:searchBox.value">
        <h3>{{dive.site}}</h3>
        <h4>{{dive.location}}</h4>
        <h2>{{dive.depth}} feet | {{dive.time}} min</h2>
      </div>
    </div>
</div>
```

Both buttons add the **(click)** attribute to the corresponding HTML element's markup. At first sight, it seems to be incompatible with the HTML standard, but it is not, "(click)" is valid HTML attribute name.

The **(click)="addDive()"** assignment instructs the Angular engine to bind the "addDive()" *template statement* of the component to the "click" event of the button. In run time, the engine identifies that the "addDive()" template statement is executed within the context of the **DiveLogComponent**, as the template file belongs to that very component. Thus, when the user clicks the Add new site button, the **addDive()** method is invoked. The method changes the **dives** member of the component, the engine detects this change and refreshes the page accordingly. The **(click)="clearDives()"** attribute follows the same logic.

This kind of binding is called *event binding* in the Angular terminology.

The Add new site button is disabled when there are no more dives left to add. The **[disabled]="!enableAdd()"** attribute defines this behavior. The right side of this assignment is a *template expression*, the left side instructs the Angular engine that the run time value of the expression should be put into the "disabled" property of the DOM element that represents the Add new site button. This syntax is called *property binding*.

> **TypeScript:** *By default, class members are public. As JavaScript does not have access modifiers, eventually, you can access even private members from component templates.*

Extracting a Type

There is still one refactoring that is worth to carry out on this little app: introducing a new type that represents dive log entries. Right now, a dive log entry is a JavaScript object, like this:

```
{
  site: 'Ponte Mahoon',
  location: 'Maehbourg, Mauritius',
  depth: 54,
  time: 38
}
```

With TypeScript in our toolbox, we could define an explicit type to describe a dive log entry. To do this refactoring, follow these steps on a copy (**Exercise-01-06**) of the previous sample (**Exercise-01-05**):

1. Add a new file, **dive-log-entry.ts** to the **app** folder:

```
export class DiveLogEntry {
  site: string;
  location: string;
  depth: number;
  time: number;

  static StockDives: DiveLogEntry[] = [
  {
    site: 'Abu Gotta Ramada',
    location: 'Hurghada, Egypt',
    depth: 72,
    time: 54
  },
  {
    site: 'Ponte Mahoon',
    location: 'Maehbourg, Mauritius',
    depth: 54,
    time: 38
  },
  {
    site: 'Molnar Cave',
    location: 'Budapest, Hungary',
    depth: 98,
    time: 62
  }];
}
```

The **DiveLogEntry** class defines an entry with four instance members, **site**, **location**, **depth**, and **time**, respectively, plus a class member, **StockDives**. The **site** and **location** members are strings, while depth and time are numbers. Evidently, **StockDives** is an array of **DiveLogEntry** items.

2. Change the definition of **dives** and **_stockDives** in `DiveLogComponent` to apply this new type:

```
import {Component} from '@angular/core';
import {DiveLogEntry} from './dive-log-entry';

@Component({
  selector: 'divelog',
  templateUrl: 'app/dive-log.template.html'
})
export class DiveLogComponent {
  public dives: DiveLogEntry[] = [];
  private _index = 0;
  private _stockDives = DiveLogEntry.StockDives;

  public enableAdd() {
    return this.index < this.stockDives.length;
  }

  public addDive() {
    if (this.enableAdd()) {
      this.dives.push(this.stockDives[this.index++]);
    }
  }

  public clearDives() {
    this.dives = [];
    this.index = 0;
  }
}
```

3. The `ContentFilterPipe` class can also leverage on `DiveLogEntry`, so change that component, too:

```
import {Pipe, PipeTransform} from '@angular/core';
import {DiveLogEntry} from './dive-log-entry';

@Pipe({name: 'contentFilter'})
export class ContentFilterPipe implements PipeTransform {
  transform(value: DiveLogEntry[], searchFor: string) : DiveLogEntry[] {
    if (!searchFor) return value;
    searchFor = searchFor.toLowerCase();
    return value.filter(dive =>
      dive.site.toLowerCase().indexOf(searchFor) >= 0 ||
      dive.location.toLowerCase().indexOf(searchFor) >= 0 ||
      dive.depth.toString().indexOf(searchFor) >= 0 ||
      dive.time.toString().indexOf(searchFor) >= 0);
  }
}
```

This refactoring does not change the compiled JavaScript files except that now the **DiveLogEntry** type is imported from its separate module. Nonetheless, it improves the developer experience a lot. Our code editor now can leverage the knowledge of the **DiveLogEntry** type. As we are typing the code, the editor can offer **DiveLogEntry** properties (Figure 1-10), and warn us about typos (Figure 1-11).

```typescript
import {Pipe, PipeTransform} from '@angular/core';
import {DiveLogEntry} from './dive-log-entry';

@Pipe({name: 'contentFilter'})
export class ContentFilterPipe implements PipeTransform {
  transform(value: DiveLogEntry[], searchFor: string) : DiveLogEntry[] {
    if (!searchFor) return value;
    searchFor = searchFor.toLowerCase();
    return value.filter(dive =>
      dive.site.toLowerCase().indexOf(searchFor) >= 0 ||
      dive.location.toLowerCase().indexOf(searchFor) >= 0 ||
      dive.depth.toString().indexOf(searchFor) >= 0 ||
      dive.|
    }
}
```
depth (property) DiveLogEntry.depth: number
location
site
time

Figure 1-10: The Visual Studio Code editor displays DiveLogEntry properties

```typescript
return value.filter(dive =>
  dive.site.toLowerCase().indexOf(searchFor) >= 0 ||
  dive.loaction.toLowerCase().indexOf(searchFor) >= 0 ||
  dive.depth.toString().indexOf(searchFor) >= 0 ||
  dive.time.toString().indexOf(searchFor) >= 0);
```

Figure 1-11: The editor recognizes the typo

Without TypeScript, we probably would not have this feature.

Summary

In this short tour, you created a static HTML page and then transformed it to a dynamic page that leverages jQuery. In contrast to the transformation that used imperative coding style, with putting Angular into the scene, you made a big step: you created a dynamic page that used no explicit code at all, but markup—plain HTML extended with Angular components, directives, and data binding expressions. In most exercises, the code was written in TypeScript.

You have only scratched the surface. In the next chapter, you will learn the tools you can use to display and style data in an Angular app.

Chapter 2: Creating Simple Applications

WHAT YOU WILL LEARN IN THIS CHAPTER

Understanding property binding and event binding

Using the `ngFor`, `ngIf`, and `ngSwitch` directives

Implementing multiple views and navigating among them

Displaying data with property binding and interpolation

Separating component responsibilities

Styling components with `ngStyle` and `ngClass`

Using style binding and class binding

In the previous chapter, we already created a tiny application that displayed dive log entries. Here, we will build on the concepts we learned and create a single page application (SPA) that shows multiple views to learn additional ways of displaying and styling data. After reading this chapter, you will understand the essential elements we can utilize to create an Angular SPA application.

Overview of the App

We are going to implement the dive site maintenance module of the Younderwater app. This module allows listing the available dive sites, adding a new site, editing or deleting an existing one. Figure 2-1 shows the maintenance page of the app.

There are many ways we can create UI for such a simple function. In this example, I use a simple approach with four views. The initial view is the list as shown in Figure 2-1, and there are three more views for adding, editing, and deleting a dive site, respectively. You may wonder why I need a separate view for removing an item. Well, I will use this to confirm the delete operation.

Figure 2-2 depicts the state-transition diagram of the app. Here, the rectangles with rounded corners represent the views. Arrows between views name the actions that change the view from one to another.

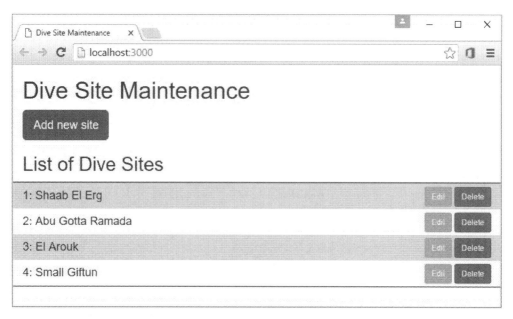

Figure 2-1: The maintenance page for managing the list of dive sites

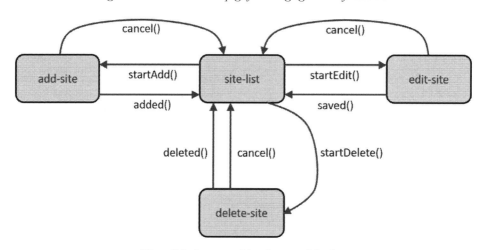

Figure 2-2: State transition diagram of the views

As this diagram suggests, actions such as **startAdd()**, **startEdit()**, and **startDelete()** are triggered with the Add new site, Edit, and Delete buttons, respectively. Each view contains a Cancel button that returns to the list without persisting any data modifications. Of course, each view contains an appropriate button to persist modifications, and these buttons trigger the **added()**, **saved()**, and **deleted()** events, respectively.

We will build this app from scratch, and add new functions step-by-step. In each phase, you will learn new concepts and features about Angular.

The app will be composed of five components. **AppComponent** is responsible for the maintenance logic. The others, **DiveListComponent**, **AddDiveComponent**, **EditDiveComponent**, and **DeleteDiveComponent** are accountable for managing the corresponding views—handling user interactions.

Creating the List View

We start building the application with the **list-site** view. The **Exercise-02-01** folder contains the result of this phase. The **index.html** file has the same structure we used earlier (Listing 2-1), it uses the **<yw-app>** element to represent the application component.

Listing 2-1: index.html (Exercise-02-01)

```
<!DOCTYPE html>
<html>
<head>
  <title>Dive Site Maintenance</title>
  <link href="/node_modules/bootstrap/dist/css/bootstrap.min.css"
    rel="stylesheet" />
  <script src="/node_modules/core-js/client/shim.min.js"></script>
  <script src="/node_modules/zone.js/dist/zone.js"></script>
  <script src="/node_modules/reflect-metadata/Reflect.js"></script>
  <script src="/node_modules/systemjs/dist/system.src.js"></script>

  <script src="systemjs.config.js"></script>
  <script>
    System.import('app').catch(function(err){ console.error(err); });
  </script>
</head>
<body>
  <yw-app>Loading...</yw-app>
</body>
</html>
```

Just like in the previous chapter, all source files related to the application logic are in the **app** folder. The **main.ts** file bootstraps **AppModule** (Listing 2-2).

Listing 2-2: app.module.ts (Exercise-02-01)

```
import {NgModule} from '@angular/core';
import {BrowserModule} from '@angular/platform-browser';

import {AppComponent} from './app.component';
import {SiteListComponent} from './site-list.component';

@NgModule({
  imports: [BrowserModule],
  declarations: [
```

```
    AppComponent,
    SiteListComponent
  ],
  bootstrap: [AppComponent]
})
export class AppModule { }
```

We do not need to touch these files; we will work only with the components that represent views, and with their HTML templates.

As Listing 2-3 shows, **AppComponent** is pretty simple. It utilizes **SiteListComponent**, and thus imports it, and adds it to its directives.

Listing 2-3: app.component.ts (Exercise-02-01)

```
import {Component} from '@angular/core';

@Component({
  selector: 'yw-app',
  template: `
    <div class="container-fluid">
      <h1>Dive Site Maintenance</h1>
      <site-list-view></site-list-view>
    </div>
})
export class AppComponent {
}
```

Here, the **@Component** annotation uses the template property to define an inline template for **AppComponent**. The template is a string literal that uses the backtick (`) character to wrap the template markup. With this construction, as you see, you can use multi-line strings. This is an ES 2015 feature—*template literal*—and TypeScript supports it.

Because the template of **AppComponent** contains a non-standard HTML markup—the **<site-list-view>** element, we have to tell the Angular engine how to process it; otherwise, Angular would render it as a simple HTML tag with no particular semantics.

We know that **SiteListComponent** manages the appearance and interactions of **<site-list-view>**. Adding **SiteListComponent** to the **declarations** metadata property in Listing 2-2 allows the engine to bind the behavior of **SiteListComponent** to this custom HTML tag.

---

*NOTE: In **AppModule**, we declared that **AppComponent** and **SiteListComponent** are both enclosed in the same module, and thus Angular can automatically make the match **<site-list-view>** with its host component, **SiteListComponent**.*

The app manages dive sites that are represented by a **DiveSite** type, as shown in Listing 2-4.

Listing 2-4: dive-site.ts (Exercise-02-01)

```
export class DiveSite {
    id: number;
    name: string;

    static FavoriteSites: DiveSite[] = [
        { id: 1, name: 'Shaab El Erg'},
        { id: 2, name: 'Abu Gotta Ramada'},
        { id: 3, name: 'El Arouk'},
        { id: 4, name: 'Small Giftun'}
    ];
}
```

The `SiteListComponent` utilizes `DiveSite` (Listing 2-5), and its `sites` property is initialized with the static `DiveSite.FavoriteSites` array.

Listing 2-5: site-list.component.ts (Exercise-02-01)

```
import {Component} from '@angular/core';
import {DiveSite} from './dive-site';

@Component({
  selector: 'site-list-view',
  templateUrl: 'app/site-list.template.html'
})
export class SiteListComponent {
  sites = DiveSite.FavoriteSites;
}
```

In contrast to `AppComponent`, `SiteListComponent` applies a template stored in an external file, `site-list.template.html`. Accordingly, the `@Component` annotation applies the `templateUrl`— and not the `template`—metadata property.

The markup of the list view is simple (Listing 2-6).

Listing 2-6: site-list.template.html (Exercise-02-01)

```
<div class="row">
  <div class="col-sm-12">
    <button class="btn btn-primary btn-lg">
      Add new site
    </button>
  </div>
</div>
<h2>List of Dive Sites</h2>
<div class="row" *ngFor="let site of sites">
  <div class="col-sm-8">
    <h4>{{site.id}}: {{site.name}}</h4>
  </div>
```

```
    <div class="col-sm-4" style="margin-top: 5px;">
      <div class="pull-right">
        <button class="btn btn-warning btn-sm">
          Edit
        </button>
        <button class="btn btn-danger btn-sm">
          Delete
        </button>
      </div>
    </div>
</div>
```

When rendering the view, Angular transforms this template. The **ngFor** directive instructs it to iterate through all elements in the object within the **sites** expression's value—an array—and put the current iteration item into the **site** local variable. The **<div>** that applies **ngFor** contains several nested elements that refer to the **site** local variable.

The **<h4>** tag's text contains two interpolations wrapped into double-curly braces:

```
<h4>{{site.id}}: {{site.name}}</h4>
```

The engine replaces these interpolations with **site.id** and **site.name**, evidently using the **id** and **name** property values of the **site** local variable.

> **NOTE**: *The leading asterisk is an essential part of the template syntax. Writing **ngFor** instead of ***ngFor** will prevent your app from running the expected way. Later in this book you will learn why we use the odd-looking syntax, and what its significance is.*

The markup adds an Edit and Delete button to each iteration. When you run the app, it displays the list of dive sites (Figure 2-3).

Creating the Add View

Now that we have successfully created the list view, it is time to implement the view that allows adding a new site to the list (**Exercise-02-02**). The **AddSiteComponent** will host this functionality, and as Listing 2-7 shows, its definition is pretty straightforward.

Listing 2-7: add-site.component.ts (Exercise-02-02)

```
import {Component} from '@angular/core';

@Component({
  selector: 'add-site-view',
  templateUrl: 'app/add-site.template.html'
})
export class AddSiteComponent {
}
```

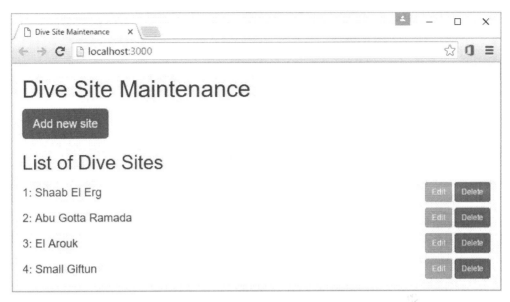

Figure 2-3: The list of dive sites

The template associated with **AddSiteComponent** adds an **<input>** and two buttons, Add and Cancel, respectively, to the markup (Listing 2-8).

Listing 2-8: add-site.template.html (Exercise-02-02)

```html
<h3>Specify the name of the dive site:</h3>
<div class="row">
  <div class="col-sm-6">
    <input class="form-control input-lg" type="text"
      placeholder="site name" />
  </div>
</div>
<div class="row" style="margin-top: 12px;">
  <div class="col-sm-6">
    <button class="btn btn-success btn">
      Add
    </button>
    <button class="btn btn-warning btn">
      Cancel
    </button>
  </div>
</div>
```

To use this new view, we need to change the template of **AppComponent**. In the previous exercise **AppComponent** had an inline template, but in this exercise, I moved it to a separate file to allow easier maintenance. While working with an inline HTML template, the code editor could not add too much support to avoid typos and other small mistakes. Moving the template to its HTML file allows the code editor to help us immediately with syntax highlighting and other context-sensitive features, such as auto-completion.

Listing 2-9 and Listing 2-10 highlight these changes.

Listing 2-9: app.component.ts (Exercise-02-02)

```
import {Component} from '@angular/core';

@Component({
  selector: 'yw-app'
  templateUrl: 'app/app.template.html',
})
export class AppComponent {
}
```

Listing 2-10: app.template.html (Exercise-02-02)

```
<div class="container-fluid">
  <h1>Dive Site Maintenance</h1>
  <site-list-view></site-list-view>
  <add-site-view></add-site-view>
</div>
```

When you run the app, the **add-site** view is displayed right beneath **list-view** (Figure 2-4).

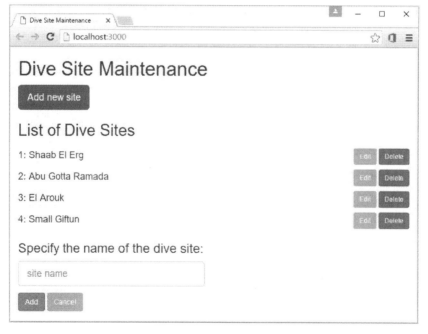

Figure 2-4: Now, this add view is visible, too

What you see is just an interim state of the app. We do not want to see both views rendered simultaneously. Instead, we intend to navigate from the list view to the add view, and back.

Navigating between the List and the Add Views

To carry on with developing the site maintenance app, we need to implement two features that we intend to leverage in the next steps. First, **AppComponent** should display only one of the views (**site-list**, **add-site**, **edit-site**, or **delete-site**) at any time. Second, we put the responsibility of modifying the list of dive sites into **AppComponent** and want to free all other components from these chores. Listing 2-11 shows the small changes in **AppComponent** that prepare these features.

Listing 2-11: app.component.ts (Exercise-02-03)

```
import {Component} from '@angular/core';
import {DiveSite} from './dive-site'

@Component({
  selector: 'yw-app',
  templateUrl: 'app/app.template.html'
})
export class AppComponent {
  sites = DiveSite.FavoriteSites.slice(0);
  currentView = 'list';

  navigateTo(view: string) {
    this.currentView = view;
  }
}
```

Here, we store the id of the view to display in the **currentView** property of **AppComponent** and provide the **navigateTo()** method to set the currently displayed view. In the previous exercise, **SiteListComponent** had the **sites** property to store the list of dive sites. Because we want to pass the responsibility of site management to **AppComponent**, we moved the **sites** property into **AppComponent** and initialized it with a clone of **DiveSite.FavoriteSites**.

To display the current view, we will change the template of **AppComponent** to use the **ngIf** Angular directive:

```
<div class="container-fluid">
  <h1>Dive Site Maintenance</h1>
  <site-list-view *ngIf="currentView == 'list'">
  </site-list-view>
  <add-site-view *ngIf="currentView == 'add'">
  </add-site-view>
</div>
```

The **ngIf** directive evaluates the expression on its right, and recreates or removes the portion of the DOM tree according to the result of this test. If the expression evaluates to true, it recreates the corresponding speck of UI; otherwise, it removes it. The template above displays the **site-list** view provided **currentView** is set to "list", or the **add-site** view, if "add" is assigned to **currentView**. Should we use another value, no views would show up.

> *NOTE: Just as you learned about the **ngFor** directive, the leading asterisk is an essential part of the template syntax, so you should not write **ngIf** instead of ***ngIf**.*

Putting **AppComponent** into the role of managing the logic means that the other components should only handle user interactions and display data. Thus, we need to carry out these tasks to implement the navigation logic between the **site-list** and **add-site** views:

1. We should pass the list of dive sites from **AppComponent** to **SiteListComponent** so that the latter one can display them.

2. We should navigate to the **add-site** view when the user clicks the Add new site button.

3. We should navigate back to **list-view** whenever the user clicks either Cancel or Add while in the **add-site** view. When Add is clicked, **AppComponent** should now the name of the new site to append to the list.

We are going to use two Angular mechanisms to implement these tasks: *input binding*, and *event binding*, respectively.

With input binding, we can pass data from a parent to a nested (child) component. Using event binding, we can raise an event from a component, and respond to that particular event in another one. Listing 2-12 shows how we can declare these bindings in the **AppComponent** template:

Listing 2-12: app.template.html (Exercise-02-03)

```
<div class="container-fluid">
  <h1>Dive Site Maintenance</h1>
  <site-list-view *ngIf="currentView == 'list'"
    [sites]="sites"
    (onAdd)="navigateTo('add')">
  </site-list-view>
  <add-site-view *ngIf="currentView == 'add'"
    (onAdded)="navigateTo('list')"
    (onCancel)="navigateTo('list')">
  </add-site-view>
</div>
```

Here, **[sites]** represent input property binding, while **(onAdd)**, **(onAdded)**, and **(onCancel)** stand for event binding. These binding types use the property binding (property name wrapped in square brackets) and event binding syntax (event name wrapped in parentheses), respectively. Soon, we will see how they become *input* and *output* bindings.

The [sites] binding uses sites as its template expression, and it means that we pass the sites property of AppComponent to SiteListView that hosts the <site-list-view> tag. When the user clicks the Add new site button in the site-list view, it will generate and event, onAdd, and due to the output binding, whenever AppComponent gets notified about this event, it will execute the navigateTo('add') template expression.

Similarly, when the user clicks the Add or Cancel view in the add-site view, these actions will raise the onAdded and onCancel events, respectively. When AppComponent receives them, it will execute the navigateTo('list') expression.

To turn property binding into *input property binding* and event binding to *output event binding*, we have to declare input and output component properties. Listing 2-13 shows how SiteListComponent carries out this task.

Listing 2-13: site-list.component.ts (Exercise-02-03)

```
import {Component, Input, Output, EventEmitter} from '@angular/core';
import {DiveSite} from './dive-site';

@Component({
  selector: 'site-list-view',
  templateUrl: 'app/site-list.template.html'
})
export class SiteListComponent {
  @Input() sites: DiveSite[];
  @Output() onAdd = new EventEmitter();

  add() {
    this.onAdd.emit(null);
  }
}
```

The component adds the @Input() annotation to sites, and @Output() to onAdd, so sites becomes an input, onAdd an output property. Output properties must be a type of EventEmitter so that they can publish events to potential listeners—this is why we initialize onAdd with a new instance of EventEmitter.

> *TypeScript: The sites property has a type declaration that marks it DiveSite[]. You can omit to declare its type without changing the app's behavior. Nonetheless, I suggest that you use it, as it adds extra information to the property and may help the readers of the code to understand the intention easier.*

The SiteListComponent template invokes the add() method (Listing 2-14) when the user clicks the Add button. This method calls the emit() method of onAdd to notify listeners that the event has occurred. We pass null to emit() to indicate that this event has no parameter to process by listeners.

Listing 2-14: site-list.template.html (Exercise-02-03)

```
<div class="row">
  <div class="col-sm-12">
    <button class="btn btn-primary btn-lg"
      (click)="add()">
      Add new site
    </button>
  </div>
</div>
<h2>List of Dive Sites</h2>
<div class="row" *ngFor="let site of sites">
  <!-- Omitted for the sake of brevity -->
</div>
```

The implementation of **AddSiteComponent** is very similar to **SiteListComponent**, but instead of one input and one output property, it contains two output properties (Listing 2-15).

Listing 2-15: add-site.component.ts (Exercise-02-03)

```
import {Component, Output, EventEmitter} from '@angular/core';

@Component({
  selector: 'add-site-view',
  templateUrl: 'app/add-site.template.html'
})
export class AddSiteComponent {
  @Output() onAdded = new EventEmitter();
  @Output() onCancel = new EventEmitter();

  added() {
    this.onAdded.emit(null);
  }

  cancel() {
    this.onCancel.emit(null);
  }
}
```

The template of **AddSiteComponent** defines click event handlers for the Add and Cancel buttons (Listing 2-16).

Listing 2-16: add-site.template.html (Exercise-02-03)

```
...
<div class="row" style="margin-top: 12px;">
  <div class="col-sm-6">
    <button class="btn btn-success btn"
      (click)="added()">
      Add
```

```
      </button>
      <button class="btn btn-warning btn"
        (click)="cancel()">
        Cancel
      </button>
    </div>
  </div>
  ...
```

Now, run the app, and try to navigate from the list to the add view. When you click the Add new site button, the app changes to the **add-site** view, as shown in Figure 2-5.

Figure 2-5: Now, we can navigate to the Add view

Implementing the Add Logic

The navigation works, but we cannot add a new site to the list because we have not implemented this logic yet. There are several ways we can do that. In the next exercise, I will follow this approach:

1. When the user clicks the Add new site button, **AppComponent** creates a new site id, passes it to **AddSiteComponent** with input property binding.

2. The user edits the new site's name and then clicks the Add button. It raises the **onAdded** event.

3. We modify the event so that it passes back the name of the new site.

4. **AppComponent** responds to the **onAdded** event and creates a new site from the id created in step 1, and the name received in the **onAdded** event.

Let's start with refactoring **AppComponent**. As Listing 2-17 shows, the component has two new methods, **addSite()**, and **siteAdded()**, respectively.

Listing 2-17: app.component.ts (Exercise-02-04)

```
import {Component} from '@angular/core';
import {DiveSite} from './dive-site'

@Component({
  selector: 'yw-app',
  templateUrl: 'app/app.template.html'
})
export class AppComponent {
  sites = DiveSite.FavoriteSites.slice(0);
  newSiteId: number;
  currentView = 'list';

  navigateTo(view: string) {
    this.currentView = view;
  }

  startAdd() {
    this.newSiteId = this.sites.map(s => s.id)
      .reduce((p, c) => p < c ? c : p) + 1;
    this.navigateTo('add');
  }

  siteAdded(newSiteName: string) {
    this.sites.push({id: this.newSiteId, name:newSiteName});
    this.navigateTo('list');
  }
}
```

The logic invokes the **startAdd()** method when the user clicks the Add new site button. When **AppComponent** receives the **onAdded** event, **siteAdd()** runs, and appends the new site to the list of existing ones.

In Listing 2-17, you can find a syntax of function definitions that may look strange:

```
s => s.id
(p, c) => p < c ? c : p
```

These are *lambda functions*, a kind of syntax sugar in TypeScript (called arrow functions in ES 2015). They are the same as if you wrote these inline functions in JavaScript:

```
function(s) { return s.id; }
function(p, c) { return p < c ? c : p; }
```

When the **startAdd()** method calculates the next site id (the maximum of ids + 1), it stores the value in the **newSiteId** property and passes it to **AddSiteComponent** with input property binding (Listing 2-18, Listing 2-19).

Listing 2-18: app.template.html (Exercise-02-04)

```html
<div class="container-fluid">
  <h1>Dive Site Maintenance</h1>
  <site-list-view *ngIf="currentView == 'list'"
    [sites]="sites"
    (onAdd)="startAdd()">
  </site-list-view>
  <add-site-view *ngIf="currentView == 'add'"
    [siteId] = "newSiteId"
    (onAdded)="siteAdded($event)"
    (onCancel)="navigateTo('list')">
  </add-site-view>
</div>
```

Observe (Listing 2-17) that **siteAdded()** accepts a string argument with the name of the new site. The **AppComponent** template passes the name that comes from the parameter of the **onAdded** event to **siteAdded()**, and **$event** represents this event parameter.

Listing 2-19: add-site.component.ts (Exercise-02-04)

```typescript
import {Component, Input, Output, EventEmitter} from '@angular/core';

@Component({
  selector: 'add-site-view',
  templateUrl: 'app/add-site.template.html'
})
export class AddSiteComponent {
  @Input() siteId: number;
  @Output() onAdded = new EventEmitter<string>();
  @Output() onCancel = new EventEmitter();
  siteName: string;

  added() {
    this.onAdded.emit(this.siteName);
  }

  cancel() {
    this.onCancel.emit(null);
  }
}
```

The component stores the new site name being edited in its **siteName** property. To pass back the new site name with the **onAdded** event, I changed **AddSiteComponent**. Instead of using an **EventEmitter** instance, I applied the generic version of this type, **EventEmitter<string>**. When the component raises the **onAdded** event, it passes the string in **siteName** to the **emit()** method, and that handles this value as the event parameter.

> ***TypeScript:*** *To allow being able to create a component that can work over a variety of types rather than a single one, TypeScript implements generic types. In the code snippet above,* `EventEmitter<T>` *is such a type, where* `T` *is a placeholder for a concrete type.* `EventEmitter<T>` *defines the* `emit()` *method to accept a parameter of type* `T`. *Thus an instance of* `EventEmitter<string>` *allows invoking* `emit()` *with a* `string` *argument.*

The last piece of the logic still missing. It is the way we allow the user to edit the new site's name. The template of `AddSiteComponent` (Listing 2-20) unravels this secret.

Listing 2-20: add-site.template.html (Exercise-02-04)

```html
<h3>Specify the name of the dive site:</h3>
<div class="row">
  <div class="col-sm-6">
    <input #siteNameBox class="form-control input-lg" type="text"
      placeholder="site name"
      (keyup)="siteName=siteNameBox.value"
      (keyup.enter)="added()" />
  </div>
</div>
<div class="row" style="margin-top: 12px;">
  <div class="col-sm-6">
    <button class="btn btn-success btn"
      (click)="added()"
      [disabled]="!siteName">
      Add
    </button>
    <button class="btn btn-warning btn"
      (click)="cancel()">
      Cancel
    </button>
  </div>
</div>
```

The `<input>` tag continuously updates the `siteName` property of `AddSiteComponent` when the user presses a key. The `#siteNameBox` attribute instructs the Angular engine to create a local variable, `siteNameBox`, which keeps a reference to the DOM representation of this very `<input>` tag. Whenever the user presses and then releases a key, the `<input>` element receives a "keyup" event from the browser. According to the declaration of the `(keyup)` attribute above, the template expression assigns the `value` property of the `<input>` element—referenced through the `siteNameBox` local variable—to the `siteName` property of `AddSiteComponent`.

Angular has a mechanism called *key event filtering*. We can create pseudo-events just for listening to a particular key. The `(keyup.enter)` attribute defines such an event, which only fires when we press the Enter key. The template expression invokes `added()`, so Enter behaves as if the user clicked the Add button.

There is one small detail to improve the user experience. When the name is empty, the Add button is disabled with the **[disabled]="!siteName"** binding.

After going through these little changes and lengthy explanations, it is time to test the app. Run the app and try adding a new site. Check that the Add button and pressing Enter works as expected.

Figure 2-6 shows the list with a newly added new dive site.

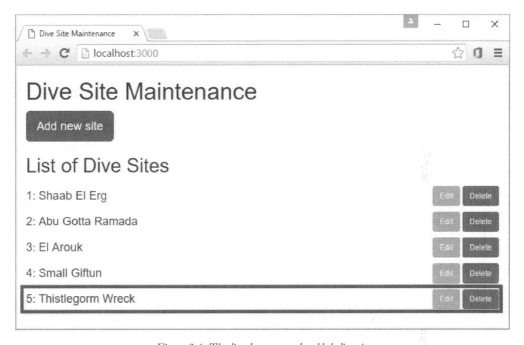

Figure 2-6: The list shows a newly added dive site

There is a little bug in the app: you can press Enter while the site name is empty, and thus you can add empty names. I guess, you can resolve this issue. In the code of the next exercise, you can check my solution in the **add-site.component.ts** file.

Editing an Item in the List

We can implement editing dive site items with the same approach we used for creating the **add-site** view. Instead of telling you every detail, I will discuss a few peculiarities of the code. You can find the complete sample of this phase in the **Exercise-02-05** folder.

As the code extract in Listing 2-21 shows, **AppComponent** utilizes the existing **currentView** property to pass the site being edited to **EditSiteComponent**.

Listing 2-21: app.component.ts (Exercise-02-05)

```
...
@Component({
  selector: 'yw-app',
  templateUrl: 'app/app.template.html'
})
export class AppComponent {
  sites = DiveSite.FavoriteSites.slice(0);
  newSiteId: number;
  currentSite: DiveSite;

  // ...
  startEdit(site: DiveSite) {
    this.currentSite = { id: site.id, name: site.name };
    this.navigateTo('edit');
  }

  siteSaved(site: DiveSite) {
    let oldSite = this.sites.filter(s => s.id == site.id)[0];
    if (oldSite) {
      oldSite.name = site.name;
    }
    this.navigateTo('list');
  }
}
...
```

If **EditSiteComponent** worked with the same dive site instance that is stored in **AppComponent**, the cancellation logic should restore the original site name value before navigating back to the list view. To avoid this plight and make things easier, **startEdit()** passes a clone of the selected item to **EditSiteComponent**.

EditSiteComponent has a very similar structure as **AddSiteComponent**, except that it utilizes a **site** input property that accepts a **DiveSite** instance, unlike **AddSiteComponent** that receives only a site id (Listing 2-22).

Listing 2-22: edit-site.component.ts (Exercise-02-05)

```
import {Component, Input, Output, EventEmitter} from '@angular/core';
import {DiveSite} from './dive-site';

@Component({
  selector: 'edit-site-view',
  templateUrl: 'app/edit-site.template.html'
})
export class EditSiteComponent {
  @Input() site: DiveSite;
  @Output() onSaved = new EventEmitter<DiveSite>();
  @Output() onCancel = new EventEmitter();
```

```
saved() {
  if (this.site.id) {
    this.onSaved.emit(this.site);
  }
}

cancel() {
  this.onCancel.emit(null);
}
}
```

The template of **EditSiteComponent** is semantically the same as the one used with **AddSiteComponent**. Of course, to allow the new **edit-site** view to work, it must be added to the template of **AppComponent** (Listing 2-23).

Listing 2-23: app.template.html (Exercise-02-05)

```html
<div class="container-fluid">
  <h1>Dive Site Maintenance</h1>
  <site-list-view *ngIf="currentView == 'list'"
    [sites]="sites"
    (onAdd)="startAdd()"
    (onEdit)="startEdit($event)">
  </site-list-view>
  <add-site-view *ngIf="currentView == 'add'"
    [siteId] = "newSiteId"
    (onAdded)="siteAdded($event)"
    (onCancel)="navigateTo('list')">
  </add-site-view>
  <edit-site-view *ngIf="currentView == 'edit'"
    [site] = "currentSite"
    (onSaved)="siteSaved($event)"
    (onCancel)="navigateTo('list')">
  </edit-site-view>
</div>
```

> *NOTE: To allow Angular bind the* **<edit-site-view>** *tag to its host component,* **EditSiteComponent**, *we have to add* **EditSiteComponent** *to the declarations of* **AppModule**.

When you run the app from the **Exercise-02-05** folder, you can edit an existing site (Figure 2-7).

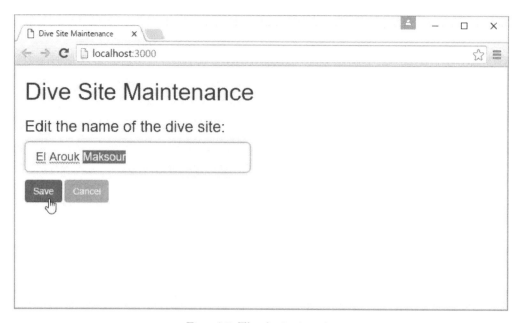

Figure 2-7: The edit view in action

Using ngSwitch

After understanding how the add and edit views were created, there is nothing surprising in the implementation of **DeleteSiteComponent** as it leverages the very same approach you already learned. You can find the app in the **Exercise-02-06** folder.

Besides implementing the new component, this app version uses another Angular directive, **ngSwitch**. In the previous exercises, **ngIf** was responsible for displaying the current view:

```
...
<site-list-view *ngIf="currentView == 'list'"
   [sites]="sites"
   (onAdd)="startAdd()"
   (onEdit)="startEdit($event)">
</site-list-view>
<add-site-view *ngIf="currentView == 'add'"
   [siteId] = "newSiteId"
   (onAdded)="siteAdded($event)"
   (onCancel)="navigateTo('list')">
</add-site-view>
<edit-site-view *ngIf="currentView == 'edit'"
   [site] = "currentSite"
   (onSaved)="siteSaved($event)"
   (onCancel)="navigateTo('list')">
</edit-site-view>
...
```

Should `currentView` have a value not handled in any `ngIf` branches, no view would show up in the page. Instead of using multiple `ngIf`, Angular offers the `ngSwitch` directive, which is more flexible—though it requires a bit lengthier markup. Listing 2-24 demonstrates how the `AppComponent` template leverages `ngSwitch` in `Exercise-02-06`.

Listing 2-24: app.template.html (Exercise-02-06)

```html
<div class="container-fluid">
  <h1>Dive Site Maintenance</h1>
  <div [ngSwitch]="currentView">
    <template ngSwitchDefault>
      <site-list-view
        [sites]="sites"
        (onAdd)="startAdd()"
        (onEdit)="startEdit($event)"
        (onDelete)="startDelete($event)">
      </site-list-view>
    </template>
    <template ngSwitchCase="add">
      <add-site-view
        [siteId] = "newSiteId"
        (onAdded)="siteAdded($event)"
        (onCancel)="navigateTo('list')">
      </add-site-view>
    </template>
    <template ngSwitchCase="edit">
      <edit-site-view
        [site] = "currentSite"
        (onSaved)="siteSaved($event)"
        (onCancel)="navigateTo('list')">
      </edit-site-view>
    </template>
    <template [ngSwitchCase]="'delete'">
      <delete-site-view
        [site] = "currentSite"
        (onDeleted)="siteDeleted($event)"
        (onCancel)="navigateTo('list')">
      </delete-site-view>
    </template>
  </div>
</div>
```

Just like `ngIf` and `ngFor`, `ngSwitch` is a structural directive, too. When you add `ngSwitch` to an HTML element, you need to declare an expression that determines the condition the value of which selects a nested template. In the listing, the outer `<div>` element says that `ngSwitch` must use the value of the `currentView` property, and the `<div>` embeds four templates. During run time, the directive selects the first one where `ngSwitchCase` matches with the condition value of `ngSwitch`. If there is no match, `ngSwitch` selects the template with `ngSwitchDefault`—provided you added such a template. Otherwise, no one is selected.

When the condition value changes, **ngSwitch** calculates the new template to display. If it is different from the current one, **ngSwitch** swaps the current with the new one.

In the listing, you can observe syntax that may seem odd at the first sight. Directive names sometimes use wrapping square brackets, at times they do not:

```
[ngSwitch]="currentView"
ngSwitchDefault
ngSwitchCase="add"
[ngSwitchCase]="'delete'"
```

The most eye-catching difference is **ngSwitchCase** that is used both with and without square brackets. So what is going on?

When Angular parses a template, it processes HTML attributes, and stores these attributes as properties of the object that represent the particular HTML element. In that object, the parser will create a property named **attr** for both the **[attr]** and **attr** attributes, however, it will change the evaluation of the attribute values.

If you use the attribute with square brackets, Angular takes the value into account as template expression and assigns the evaluated to the appropriate property. Thus the processing of **[ngSwitch]="currentView"** results in storing the value of **currentView** in the **ngSwitch** property of the **<div>** tag's in-memory representation.

If you use an attribute value without the square brackets, Angular assigns the attribute's value as a string to the corresponding property. When parsing the **ngSwitchCase="add"** markup, the engine assigns the "add" string value to the **ngSwitchCase** property. When Angular interprets the **[ngSwitchCase]="'delete'"** markup, it sets **ngSwitchCase** to the "delete" string after evaluating the attribute's value as an expression—as it is set to the literal string "delete". If you wrote **[ngSwitchCase]="delete"** (without single quotes), the engine would assign the value of the **delete** expression to **ngSwitchCase**, which would be undefined.

> **NOTE**: *Later in the book, you will learn all details behind property binding.*

In Listing 2-24, I wrapped all switch cases into the **<template>** tag deliberately. We can use another syntax with the "*" prefix, similarly as you did with ***ngIf**. So, we could have written this:

```
<div [ngSwitch]="currentView">
  <site-list-view *ngSwitchDefault
    [sites]="sites"
    (onAdd)="startAdd()"
    (onEdit)="startEdit($event)"
    (onDelete)="startDelete($event)">
  </site-list-view>
  <add-site-view *ngSwitchCase="'add'"
    [siteId] = "newSiteId"
    (onAdded)="siteAdded($event)"
    (onCancel)="navigateTo('list')">
```

```
  </add-site-view>
  <edit-site-view *ngSwitchCase="'edit'"
    [site] = "currentSite"
    (onSaved)="siteSaved($event)"
    (onCancel)="navigateTo('list')">
  </edit-site-view>
  <delete-site-view *ngSwitchCase="'delete'"
    [site] = "currentSite"
    (onDeleted)="siteDeleted($event)"
    (onCancel)="navigateTo('list')">
  </delete-site-view>
</div>
```

Behind the scenes, Angular transforms the "*" syntax transforms to the one we used in Listing 2-24. In *Chapter 5, The <template> Element*, you will learn more details about this syntax sugar.

Styling Components

CSS adds many values to web apps. With the modern CSS3, you can implement complex tasks with stylesheets—even without JavaScript. Angular provides great styling support, as you will learn in this section.

The framework offers two built-in directives, **ngStyle** and **ngClass**, which allow you to manage the **style** and **class** HTML attributes in a programmatic way. You can define styles that work in the context of the component that declares them, but they do not influence the appearance of other ones.

Using ngStyle

Exercise 02-07 contains a modified **SiteListComponent** that sets alternating background color for the items of the dive site list (Figure 2-8).

The component template has only slight changes; it applies the **ngStyle** attribute (Listing 2-25).

Listing 2-25: site-list.template.html (Exercise-02-07)

```
...
<div class="row" *ngFor="let site of sites; let i=index"
  [ngStyle]="{ 'background-color': i%2 == 0 ? '#dddddd' : 'inherit'}">
  <div class="col-sm-8">
    <h4>{{site.id}}: {{site.name}}</h4>
  </div>
  <div class="col-sm-4" style="margin-top: 5px;">
    <div class="pull-right">
      <button class="btn btn-warning btn-sm"
        (click)="edit(site)">
        Edit
      </button>
```

```
        <button class="btn btn-danger btn-sm"
          (click)="delete(site)">
          Delete
        </button>
      </div>
    </div>
</div>
```

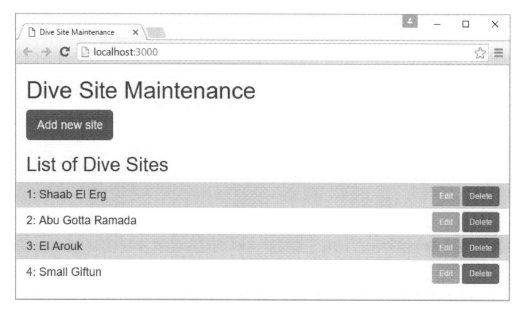

Figure 2-8: The dive site list with alternating background

In this listing, **ngStyle** is applied on each list row. The directive expects an object, where the object's keys (property names) are taken into account as style properties, the object values are the values of the corresponding style properties. The highlighted expression sets the **background-color** style property to "#dddddd" for rows with an even index.

Observe, the **ngStyle** expression uses a feature of the **ngFor** directive, namely that **ngFor** provides several exported values, such as **index** in the markup snippet above. To use any of these values, they need to be aliased to local variables, as the sample does. With the "**let i=index**" expression—it is a part of the **ngFor** *microsyntax*, and not the Angular template expression syntax—you can access the current loop index in the local variable, **i**.

> *NOTE: You will learn more details about **ngFor** and microsyntaxes later in this book.*

You can make this code more straightforward with the **even** value of **ngFor**, which is set to true if the current item has an even index:

```
<div class="row" *ngFor="let site of sites; let e=even"
  [ngStyle]="{ 'background-color': e ? '#dddddd' : 'inherit'}">
  <!-- ... -->
</div>
```

Using Style Binding

Angular implements a very convenient syntax, style binding, which makes setting style properties even easier and more readable. You can use the **[style.*style-property*]** attribute name to set a single style value. The **Exercise-02-08** folder contains the modified version of **Exercise-02-07**, and it uses this way to set the alternating background color:

```
<div class="row" *ngFor="let site of sites; let i=index"
  [style.background-color]="i%2 == 0 ? '#dddddd' : 'inherit'">
  <div class="col-sm-8">
    <h4 [style.color]="'maroon'">{{site.id}}: {{site.name}}</h4>
  </div>  <!-- ... -->
</div>
```

Take care when using style binding! When you want to set a literal string value, do not forget about the single quotes; otherwise the value you intended to be a literal is evaluated as an expression, and it may lead to errors or unexpected results. In the code snippet above, the text color of the **<h4>** tag is set to maroon with a string literal.

You can use non-existing style property names, with no warning or error raised. For example, if you used **[style.colour]** instead of the correct **[style.color]**, the text color would not be set, but the app would carry on without any error raised or message logged.

Using ngClass

When creating single page applications, web developers often add and remove CSS classes dynamically. The **ngClass** directive makes this task easier. Similarly to **ngStyle**, the template expression of **ngClass** is taken into account as an object: keys (property names) represent CSS class names, property values are evaluated as Boolean expressions. Where a value is true, the corresponding key is added to the CSS classes. If a value evaluates to false, the particular key is removed from the CSS class list. For example, the following **ngClass** directive adds the **square** and **rectangle** CSS classes to the **<div>** element, and removes the **circle** class:

```
<div [ngClass]="{ square: 3 > 2, circle: '', rectangle: true }">
  This tag uses ngClass
</div>
```

Angular offers *class binding*, a simplified syntax to add or remove a single CSS class. You can use the **[class.*class-name*]** attribute name to name a class. If the template expression of the attribute evaluates to true, the class is added; otherwise, it is removed.

The sample in the **Exercise-02-09** folder demonstrates **ngClass** and class binding. The template of the **SiteListComponent** (Listing 2-26) uses these features to add some styling to the list of dive sites.

Listing 2-26: site-list.template.html (Exercise-02-09)

```
...
<h2>List of Dive Sites</h2>
<div class="row" *ngFor="let site of sites; let e=even; let f=first; let
l=last"
  [ngClass]="{ evenRow: e }"
  [class.topRow]="f"
  [class.bottomRow]="l">
  <div class="col-sm-8">
    <h4>{{site.id}}: {{site.name}}</h4>
  </div>
  <!-- -->
</div>
```

This markup leverages the **even**, **first**, and **last** exported values of **ngFor**, which all are Booleans values. You already utilized **even**. As their names suggest, **first** indicates whether the current item is the first one in the iteration, **last** indicates the last one. These flags are assigned to the **e**, **f**, and **l** local variables, and referenced in the **[ngClass]**, **[class.topRow]**, and **[class.borttomRow]** attributes. Depending on the current iteration values of these flags, the **evenRow**, **topRow**, and **bottomRow** classes are added to the current style, or removed from there.

When you run the sample, you can observe the effect of these styles as shown in Figure 2-9.

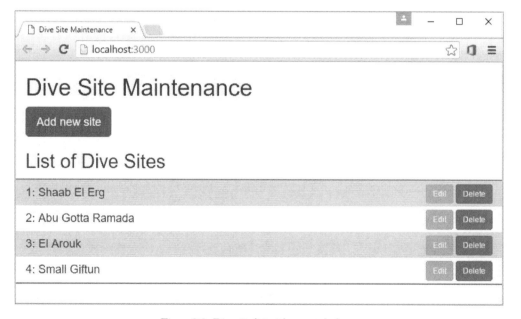

Figure 2-9: Dive site list with new style classes

Style Shims

Usually, you would put CSS classes in separate style sheet files, and would load them in the `<head>` section's `<link>` elements. If you had many styles and components, you have to make extra efforts to keep them easily maintainable. Collecting them into a single `.css` file, or splitting them into a few ones would not help.

With Angular, you can place style definitions into the components that apply them. Listing 2-27 shows how `SiteListComponent` declares the `evenRow`, `topRow`, and `bottomRow` style classes.

Listing 2-27: site-list.component.ts (Exercise-02-09)

```
import {Component, Input, Output, EventEmitter} from '@angular/core';
import {DiveSite} from './dive-site';

@Component({
  selector: 'site-list-view',
  templateUrl: 'app/site-list.template.html',
  styles: [`
    .evenRow {
      background-color: #dddddd;
    }

    .topRow {
      border-top: 2px solid #808080;
    }

    .bottomRow {
      border-bottom: 2px solid #808080;
    }
  `]
})
export class SiteListComponent {
  // --- Omitted for the sake of brevity
}
```

The `styles` component metadata property accepts an array of strings; each string can define one or more CSS rules. When you apply the `styles` metadata property, Angular uses style shims: with a tricky mechanism, those styles are applied only to the particular component.

The sample in the `Exercise-02-09` folder demonstrates it. Both the `AddSiteComponent` and `EditSiteComponent` templates contain `<h3>` elements:

```
<!-- AddSiteComponent template -->
<h3>Specify the name of the dive site:</h3>
...

<!-- EditSiteComponent template -->
<h3>Edit the name of the dive site:</h3>
...
```

AddSiteComponent defines a style rule for **<h3>**:

```
// --- AddSiteComponent
@Component({
  selector: 'add-site-view',
  templateUrl: 'app/add-site.template.html',
  styles: [`
    h3 {
      font-weight: bold;
      color: maroon;
    }
  `]
})
export class AddSiteComponent {
  // ...
}
...

// --- EditSiteComponent
@Component({
  selector: 'edit-site-view',
  templateUrl: 'app/edit-site.template.html'
})
export class EditSiteComponent {
  // ...
}
```

Due to style shims, the **<h3>** style rule in **AddSiteComponent** does not have any effect in the template of **EditSiteComponent**. When you run the app, you can observe this feature (Figure 2-10, Figure 2-11).

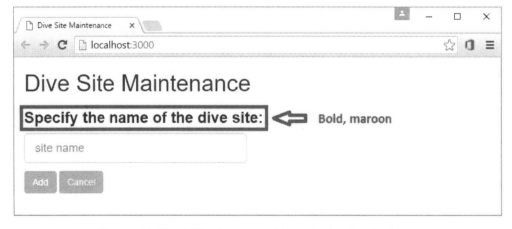

Figure 2-10: The AddSiteComponent <h3> style rule affects the add view

Figure 2-11: The AddSiteComponent <h3> style rule does not affect the edit view

Summary

With structural directives, such as **ngFor**, **ngIf**, and **ngSwitch**, you can repeat, add (display), or remove (hide) template parts conditionally.

It is easy to display data with interpolation: you can wrap template expression between double curly braces—Angular will evaluate them during run time.

Your apps contain more than one components, most often they compose parent-child or parent-children relationships. You can pass down information to children through input property binding. Child components can emit events to notify their parents.

Angular binds values to the properties of HTML elements in-memory DOM representation. You can use the literal binding syntax (**attr="literal value"**) to assign string literals to properties, or the property binding syntax (**[attr]="template expression"**) to let Angular evaluate the template expression and assign the result to the particular property.

To style your components, you can use the **ngStyle** and **ngClass** directives, style binding and class binding.

In the next chapter, you will learn other Angular concepts that make your apps more modular.

Chapter 3: Using Powerful Angular Concepts

WHAT YOU WILL LEARN IN THIS CHAPTER

Understanding what services are

Changing components to interact through a service

Configuring Angular dependency injection

Creating attribute directives

Getting an overview of Angular routing

Refactoring a multiple-view app to use the component router

In the previous chapter, we built a single page application (dive site maintenance), which utilized four views. The logic of the application was handled by **AppComponent**, and this very same component was responsible for navigation among the views. The application was not complex, but we had to add several code snippets just for the sake of passing information between components.

Angular provides great concepts that may help us create that app in a more straightforward way. In this chapter, you will learn about them.

Using Services

The dive site maintenance app—as we finished it at the end of the previous chapter—uses five components: **SiteListComponent**, **AddSiteComponent**, **EditSiteComponent**, **DeleteSiteComponent**, and **AppComponent**. The first four of them implement view logic—accepting input and raising navigation events—, while **AppComponent** encapsulates the logic of managing the dive site list, and navigation. Through events, the view components notify **AppComponent** about user interactions that affect navigation.

In this pattern, by means of responsibility, **AppComponent** is a bit overloaded. What if we could extract to dive site maintenance logic into a separate object, and all other components could leverage on that object to carry out their tasks?

Angular offers such a concept. A *service* is a class with a clear responsibility, a narrow set of operations proffered to other components. Let's improve the dive site maintenance app with services!

Changing Component Interactions

In the previous chapter, components implemented the collaboration pattern shown by the state-transition diagram in Figure 3-1.

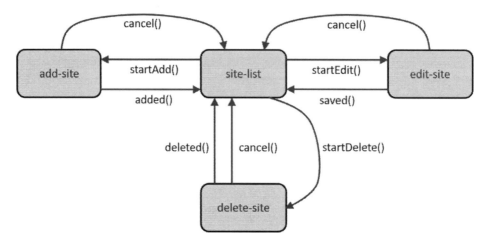

Figure 3-1: The state transition diagram of the app

According to the state names, it is easy to identify the components that manage a particular state. However, you cannot see anything that would represent **AppComponent**. Well, it is everywhere! Each arrow that represents a state transition goes through **AppComponent**. A couple of state transitions, **added()**, **saved()**, and **deleted()**, respectively, have side effects: they change the list of dive sites.

With a service, we can apply another pattern to implement state transitions (Figure 3-2). Here, each view saves any dive site information changes by invoking an appropriate service operation and navigates back to the **site-list** view. For example, if the user clicks either the Add or Cancel button in the **add-site** view, the **close()** operation returns to the **site-list** view. Nonetheless, the Add button invokes the **addSite()** service operation before closing the view. By the time any view navigates back to **site-list**, the list of dive sites may have been changed. Thus, **site-list** refreshes its content by calling the **getAllSites()** service operation.

Creating the Service

The **Exercise-03-01** folder contains the first phase of refactoring the dive site management application. The biggest change is that it introduces a new class, **SiteManagementService** (Listing 3-1).

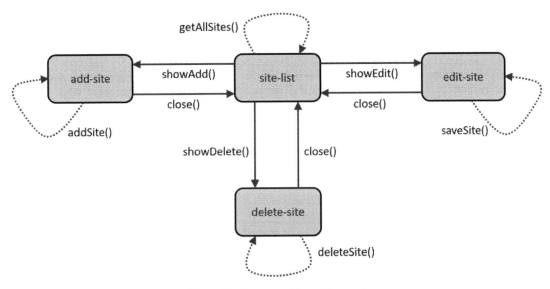

Figure 3-2: State transitions with a service

Listing 3-1: site-management.service.ts (Exercise-03-01)

```
import {Injectable} from '@angular/core';
import {DiveSite} from './dive-site';

@Injectable()
export class SiteManagementService {
  private _sites = DiveSite.FavoriteSites.slice(0);

  getAllSites(): DiveSite[] {
    return this._sites.slice(0);
  }

  getSiteById(id: number): DiveSite {
    let site = this._sites.filter(s => s.id == id)[0];
    return site
      ? {id: site.id, name: site.name }
      : null;
  }

  addSite(newSite: DiveSite) {
    newSite.id = this._sites.map(s => s.id)
      .reduce((p, c) => p < c ? c : p) + 1;
    this._sites.push(newSite);
  }

  saveSite(site: DiveSite) {
    let oldSite = this._sites.filter(s => s.id == site.id)[0];
    if (oldSite) {
      oldSite.name = site.name;
    }
```

```
      }

    deleteSite(id: number) {
      let oldSite = this._sites.filter(s => s.id == id)[0];
      if (oldSite) {
        let siteIndex = this._sites.indexOf(oldSite);
        if (siteIndex >= 0) {
          this._sites.splice(siteIndex, 1);
        }
      }
    }
  }
}
```

As you will learn soon, Angular uses its dependency injection framework to pass services to other components. This framework leverages on class metadata, and the **@Injectable()** decorator provides this metadata. Right now, it is enough to notice that you need to add this decoration to a service class.

The **SiteManagementService** class implementation is pretty straightforward. We store the current list of dive sites in the **_sites** private field and initialize it with a clone of **DiveSite.FavoriteSites** just to have a few sites in the list right at the beginning for demonstration purposes.

The service is an autonomous entity that handles its data separately from the external world. Each dive site is a JavaScript object, so if we would pass a reference to a dive site to an external entity, and that could change any dive site property. Thus, such a modification would override the dive site stored within the service. We must not allow it, so we always pass back cloned instances. The two methods, **getAllSites()**, and **getSiteById()** follows this approach.

Having **SiteManagementService** in our hands, the code of **AppComponent** gets pretty simple (Listing 3-2).

Listing 3-2: app.component.ts (Exercise-03-01)

```
import {Component} from '@angular/core';
import {SiteListComponent} from './site-list.component';

@Component({
  selector: 'yw-app',
  templateUrl: 'app/app.template.html'
})
export class AppComponent {
  siteId: number;
  currentView = 'list';

  navigateTo(view: string) {
    this.currentView = view;
  }
}
```

With the help of the **currentView** property and the **navigateTo()** method, **AppComponent** is able to navigate to other views, just like it was in the previous chapter. It provides a **siteId** property so that the edit and delete views can be notified about the dive site they should operate on.

In Listing 3-3, you can check that the template of **AppComponent** utilizes only **siteId** and **navigateTo()** to implement its navigation logic.

Listing 3-3: app.template.html (Exercise-03-01)

```html
<div class="container-fluid">
  <h1>Dive Site Maintenance</h1>
  <div [ngSwitch]="currentView">
    <site-list-view *ngSwitchDefault
      (onAdd)="navigateTo('add')"
      (onEdit)="siteId=$event; navigateTo('edit')"
      (onDelete)="siteId=$event; navigateTo('delete')">
    </site-list-view>
    <add-site-view *ngSwitchCase="'add'">
    </add-site-view>
    <edit-site-view *ngSwitchCase="'edit'">
    </edit-site-view>
    <delete-site-view *ngSwitchCase="'delete'">
    </delete-site-view>
  </div>
</div>
```

You can observe, too, that event properties can be bound to inline statement sets—following the JavaScript syntax—where statements are separated by semicolons. Be aware that these expressions are parsed by Angular, so you cannot use everything that is at your disposal in JavaScript.

When you run the app, it successfully bootstraps **AppComponent** and navigates to the list view (Figure 3-3). However, it does not display any dive sites because **AppComponent** does not pass the list of them to **SiteListComponent**.

Figure 3-3: The list view does not display dive sites

It is time to incorporate **SiteManagementService** into our code!

Injecting the Service into Components

Angular uses class constructors to inject a service into a class. Listing 3-4 shows the changes that prepare **SiteListComponent** to work with **SiteManagementService**.

Listing 3-4: app.template.html (Exercise-03-02)

```
import {Component, Output, EventEmitter} from '@angular/core';
import {DiveSite} from './dive-site';
import {SiteManagementService} from './site-management.service'

@Component({
  selector: 'site-list-view',
  templateUrl: 'app/site-list.template.html',
  // --- 'styles' omitted for the sake of brevity
})
export class SiteListComponent {
  sites: DiveSite[];
  @Output() onAdd = new EventEmitter();
  @Output() onEdit = new EventEmitter<number>();
  @Output() onDelete = new EventEmitter<number>();

  constructor(private siteService: SiteManagementService) {
    this.sites = siteService.getAllSites();
  }

  add() {
    this.onAdd.emit(null);
  }

  edit(siteId: number) {
    this.onEdit.emit(siteId);
  }

  delete(siteId: number) {
    this.onDelete.emit(siteId);
  }
}
```

The code defines a class constructor—that can accept zero, one, or more arguments— with the **constructor** keyword. In this listing, the constructor takes a single argument, **siteService**, which is declared as type of **SiteManagementService**. When a **SiteListComponent** object is instantiated, in a magical way it automatically gets a **SiteManagementService** instance. The constructor uses this very instance to query the list of dive sites—it invokes the **getAllSites()** service operation. The template of **SiteListComponent** displays these dive sites.

> **TypeScript**: *Class constructors support marking constructor parameters as private or public. If a parameter is signed as* **private** *or* **public**, *the compiler automatically creates a private or public field, respectively, and stores the parameter value in that member. Within class method bodies, this field can be accessed. So, if you need, you can access the* **siteService** *parameter accepted by the constructor as* **this.siteService** *in any method's body.*

The list view raises three events; the component represents them by the **onAdd**, **onEdit**, and **onDelete** output properties. While **onEdit** and **onDelete** utilize an event parameter to pass the id of the affected site, **onAdd** does not use any. To adjust the template to these tiny changes, we need to carry out only subtle modifications as highlighted in Listing 3-5.

Listing 3-5: site-list.template.html (Exercise-03-02)

```html
<div class="row">
  <div class="col-sm-12">
    <button class="btn btn-primary btn-lg"
      (click)="add()">
      Add new site
    </button>
  </div>
</div>
<h2>List of Dive Sites</h2>
<div class="row" *ngFor="let site of sites; let e=even; let f=first; let
l=last"
  [ngClass]="{ evenRow: e }"
  [class.topRow]="f"
  [class.bottomRow]="l">
  <div class="col-sm-8">
    <h4>{{site.id}}: {{site.name}}</h4>
  </div>
  <div class="col-sm-4" style="margin-top: 5px;">
    <div class="pull-right">
      <button class="btn btn-warning btn-sm"
        (click)="edit(site.id)">
        Edit
      </button>
      <button class="btn btn-danger btn-sm"
        (click)="delete(site.id)">
        Delete
      </button>
    </div>
  </div>
</div>
```

We are very close to kicking off the new **SiteListComponent** and see it in action. Nonetheless, we need one more step so that the magical service injection step of Angular could work.

Add the highlighted changes to metadata property to the **@Component()** annotation of
AppComponent:

```
import {Component} from '@angular/core';
import {SiteManagementService} from './site-management.service';

@Component({
  selector: 'yw-app',
  templateUrl: 'app/app.template.html',
  providers: [SiteManagementService]
})
export class AppComponent {
  siteId: number;
  currentView = 'list';

  navigateTo(view: string) {
    this.currentView = view;
  }
}
```

This small change bounds our components with **SiteManagementService** and ensures that a
service instance will be injected to each of them requesting it. Now the list view works again—but
this time, it leverages the service (Figure 3-4).

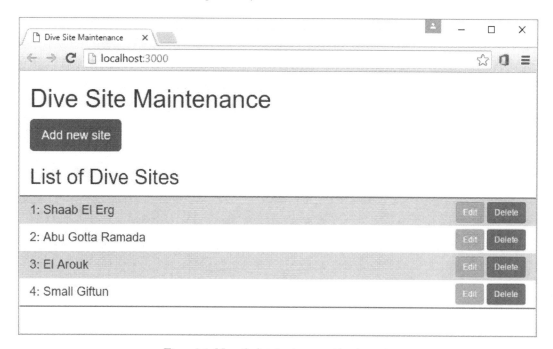

Figure 3-4: Now the list view is powered by the service

Angular Dependency Injection

Let's dive I bit deeper into this mechanism, the Angular dependency injection!

The **SiteListComponent** class uses **SiteManagementService** to implement its operation. We can say that **SiteListComponent** depends on **SiteManagementService**. We could create a **SiteManagementService** instance within **SiteListComponent** and utilize it this way:

```
export class SiteListComponent {
  sites: DiveSite[];
  siteService = new SiteManagementService();
  @Output() onAdd = new EventEmitter();
  @Output() onEdit = new EventEmitter<number>();
  @Output() onDelete = new EventEmitter<number>();

  constructor() {
    this.sites = this.siteService.getAllSites();
  }
  // ...
}
```

Well, this approach has two serious issues. First, **SiteListComponent** and **SiteManagementService** are tightly-coupled, and that makes testing of **SiteListComponent** hard. Second, our application logic assumes that we utilize the very same **SiteManagementService** in all components, including **AddSiteComponent**, **EditSiteComponent**, and **DeleteSiteComponent**. The only way to ensure this condition is to create a singleton **SiteManagementService** instance and pass it to the component instances.

Thus we need to use a pattern, known as dependency injection, to pass dependency objects to their consumers. This is what we had done when we created the **SiteListComponent** constructor:

```
export class SiteListComponent {
  // ...
  constructor(private siteService: SiteManagementService) {
    this.sites = siteService.getAllSites();
  }
  // ...
}
```

When our application is about to display the UI of the list view, it needs to instantiate **SiteListComponent**. Proper instantiation requires a **SiteManagementService** instance. Otherwise, the constructor would not work.

This job is assigned to an *injector*. When Angular needs to work with an instance of a particular class, such as a component, a pipe, a service—or with another kind of essential object—, the framework turns to the injector: "give me an instance of **ComponentA**!"

The injector has its lookup information—based on the metadata we decorate our classes with—, and finds out how **ComponentA** should be instantiated. This lookup information is represented by a *provider*. If **ComponentA** depends on other classes or objects, the injector recursively gets them, taking care of resolving the dependencies of dependencies, too. In many cases—just like in our dive site management app—, we do not want to create new instances of dependencies, but rather using a singleton instance.

In reality, every component has an injector. It does not mean that each component has its own dedicated one, components may share the same instance. When the application bootstraps a component, Angular creates an injector for that one. The child components of the application also have their injectors.

When Angular needs to resolve a dependency for a certain component, the framework turns to the injector. If the component's injector lacks the provider—we can say that the injector has no dependency lookup information—, or it cannot resolve the request, the framework passes the request up the injector of the parent component. This mechanism goes on until the dependency can be satisfied or there is no more injector to pass the request. In the latter case, Angular throws an error.

Now, it is enough from dry theory. Let's see, how dependency injection works in the dive site management application! We have a hierarchy of two components, each of which has its injector (Figure 3-5).

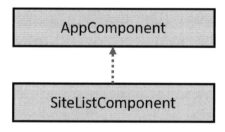

Figure 3-5: Injector hierarchy

In **main.ts**, the app bootstraps **AppComponent**, and it gets its injector. Because the template of **AppComponent** involves the highlighted markup that results in creating an instance of **SiteListComponent**, there is a parent-child component relationship between them:

```
...
<!-- AppComponent template -->
<div [ngSwitch]="currentView">
  <site-list-view *ngSwitchDefault
    (onAdd)="navigateTo('add')"
    (onEdit)="siteId=$event; navigateTo('edit')"
    (onDelete)="siteId=$event; navigateTo('delete')">
  </site-list-view>
  <!-- ... -->
</div>
...
```

Earlier I told you that there is a single line that plays a key role in starting our app working properly:

```
...
@Component({
  selector: 'yw-app',
  templateUrl: 'app/app.template.html',
  directives: [
      SiteListComponent,
      AddSiteComponent,
      EditSiteComponent,
      DeleteSiteComponent
  ],
  providers: [SiteManagementService]
})
export class AppComponent {
  // ...
}
```

This line creates a provider in the injector that belongs to **AppComponent** (Figure 3-6). Angular interprets this kind of provider declaration—of course, there are other types of provider declarations—that if a **SiteManagementService** dependency is requested, then a lazy-created **SiteManagementService** instance should be retrieved.

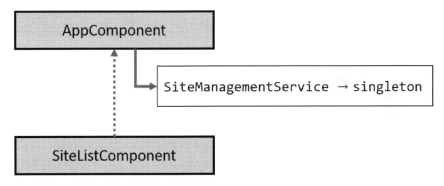

Figure 3-6: Dependency injection hierarchy with a provider

When Angular is about to create a **SiteListComponent** instance, it needs a **SiteManagementService** object, so the dependency injection module starts the resolution process with the injector of **SiteListComponent**. Because this injector does not have any provider, the resolution goes on with the **AppComponent** injector. There is a provider for **SiteManagementService**. The first time this provider is applied, Angular creates a new **SiteManagementService**. The resolution process completes successfully, and Angular creates a new instance of the **SiteListComponent** injecting the newly created **SiteManagementService** instance into the constructor.

The next time when Angular needs to instantiate **SiteListComponent**, the dependency resolution takes the same way. However, that time the framework does not create a new **SiteManagementService** instance, it retrieves and reuses the previously generated one.

Refactoring AddSiteComponent

We can follow the approach utilized in **SiteListComponent** to refactor the other components. Listing 3-6 shows how **AddSiteComponent** leverages dependency injection.

Listing 3-6: add-site.component.ts (Exercise-03-03)

```
import {Component, Output, EventEmitter} from '@angular/core';
import {SiteManagementService} from './site-management.service'

@Component({
  selector: 'add-site-view',
  templateUrl: 'app/add-site.template.html'
})
export class AddSiteComponent {
  @Output() onClosed = new EventEmitter();
  siteName: string;

  constructor(private siteService: SiteManagementService) {
  }

  add() {
    this.siteService.addSite({id: 0, name:this.siteName});
    this.onClosed.emit(null);
  }

  cancel() {
    this.onClosed.emit(null);
  }
}
```

The code builds on the TypeScript feature that private constructor parameters are automatically mapped to private class members, namely the compiler automatically generates a **siteService** field.

To let the Save and Cancel button return to the list view, we need to modify the **AppComponent** template:

```
...
<add-site-view *ngSwitchCase="'add'"
  (onClosed)="navigateTo('list')">
</add-site-view>
...
```

The app works correctly only if the same `SiteManagementService` instance is passed both to `SiteListComponent` and `AddSiteComponent`. Because both of these classes become children of `AppComponent`, the framework will inject the very same service instance into them. Figure 3-7 shows the injector hierarchy when Angular is about to instantiate `AddSiteComponent` the first time.

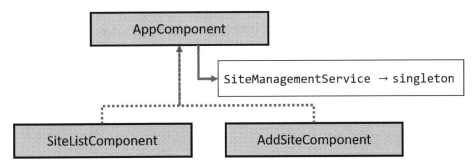

Figure 3-7: Component and injector hierarchy

According to the application logic, there is always a `SiteListComponent` already created by the time `AddSiteComponent` is to be instantiated. Thus, the dependency resolution mechanism will find a cached instance of `SiteManagementService`, which was injected to `SiteListComponent` earlier, and it will inject the very same instance into `AddSiteComponent`, too.

Of course, the template of `AddSiteComponent` is slightly changed as shown in Listing 3-7.

Listing 3-7: add-site.template.html (Exercise-03-03)

```html
<h3>Specify the name of the dive site:</h3>
<div class="row">
  <div class="col-sm-6">
    <input #siteNameBox class="form-control input-lg" type="text"
      placeholder="site name"
      (keyup)="siteName=siteNameBox.value"
      (keyup.enter)="add()" />
  </div>
</div>
<div class="row" style="margin-top: 12px;">
  <div class="col-sm-6">
    <button class="btn btn-success btn"
      (click)="add()"
      [disabled]="!siteName">
      Add
    </button>
    <button class="btn btn-warning btn"
      (click)="cancel()">
      Cancel
    </button>
  </div>
</div>
```

Refactoring EditSiteComponent

Now, that you have understood the fundamentals of the Angular dependency injection mechanism, it is easy to comprehend how the refactored `EditSiteComponent` works (Listing 3-8).

Listing 3-8: edit-site.component.ts (Exercise-03-04)

```
import {Component, Input, Output, EventEmitter} from '@angular/core';
import {SiteManagementService} from './site-management.service'

@Component({
  selector: 'edit-site-view',
  templateUrl: 'app/edit-site.template.html'
})
export class EditSiteComponent {
  @Input() siteId: number;
  siteName: string;

  @Output() onClosed = new EventEmitter();

  constructor(private siteService: SiteManagementService) {
    this.siteName = this.siteService
      .getSiteById(this.siteId).name;
  }

  save() {
    this.siteService.saveSite({id: this.siteId, name:this.siteName});
    this.onClosed.emit(null);
  }

  cancel() {
    this.onClosed.emit(null);
  }
}
```

This view allows editing the name of the dive site. The component stores the name to be edited in its `siteName` field. When the constructor receives the `SiteManagementService`, it invokes the `getSiteById()` operation to query the particular dive site information, and stores the site name.

In the `AppComponent` template, we modify `<edit-site-view>` to pass the site's id, and be able to return back to the list view:

```
<edit-site-view *ngSwitchCase="'edit'"
  [siteId]="siteId"
  (onClosed)="navigateTo('list')">
</edit-site-view>
```

Unfortunately, this approach does not work. When you run the app and start editing any site, the edit view remains empty, and the console output contains error messages (Figure 3-8).

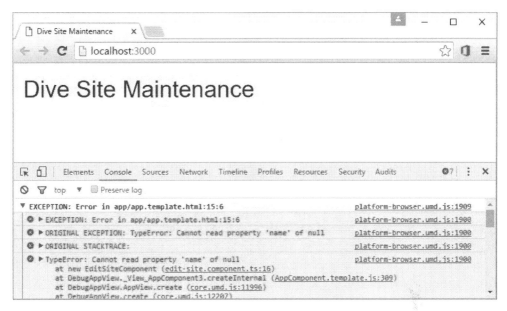

Figure 3-8: The edit view does not work.

The reason of the error is simple, but not easy to find for the first try. The id of the site to edit is passed to **EditSiteComponent** with input property binding from **AppComponent**:

```
...
<edit-site-view
  [siteId]="siteId"
  (onClosed)="navigateTo('list')">
</edit-site-view>
...
```

In **EditSiteComponent** constructor we use this **siteId** to initialize the site name to edit:

```
...
export class EditSiteComponent {
  @Input() siteId: number;
  siteName: string;

  constructor(private siteService: SiteManagementService) {
    this.siteName = this.siteService
      .getSiteById(this.siteId).name;
  }
  // ...
}
```

And this is when the error comes from: by the time the constructor executes, the **siteId** input property is unset. If we think it over, it is entirely logical. First, Angular needs to create an **EditSiteComponent** instance so that it can set its **siteId** input property. Thus, when the constructor runs, it does not have a value for **siteId** yet.

We need to initialize the `siteName` field when we are sure that `siteId` has already been set. There are two easy ways to do that.

Using a Property Setter

TypeScript allows as to create class properties with getter and setter methods—leveraging on JavaScript features. To catch the moment when `siteId` is set, we create a property setter for `siteId` and initialize `siteName` there. This solution is simple, as Listing 3-9 shows.

Listing 3-9: edit-site.component.ts (Exercise-03-05)

```
import {Component, Input, Output, EventEmitter} from '@angular/core';
import {SiteManagementService} from './site-management.service'

@Component({
  selector: 'edit-site-view',
  templateUrl: 'app/edit-site.template.html'
})
export class EditSiteComponent {
  private _siteId: number;
  siteName: string;

  @Input() set siteId(id: number) {
    this._siteId = id;
    this.siteName = this.siteService.getSiteById(id).name;
  }

  @Output() onClosed = new EventEmitter();

  constructor(private siteService: SiteManagementService) {
  }

  save() {
    this.siteService.saveSite({id: this._siteId, name:this.siteName});
    this.onClosed.emit(null);
  }

  cancel() {
    this.onClosed.emit(null);
  }
}
```

As the code shows, we store the site id in the `_siteId` backing field. The `siteId` property setter not only assigns the property value to `_siteId` but also queries the site information from the service.

> ***TypeScript***: *The syntax we used above is how TypeScript declares a property setter. If we had intended to create a get setter, we had used the* `get siteId() { return _siteId }` *definition.*

Using the OnChanges Lifecycle Hook

There is an alternative solution. Instead of an input property setter, we can use the **OnChanges** lifecycle hook—this is how I implemented **DeleteSiteComponent**.

Angular manages the entire lifecycle of components from their creation to destroying and removing them from the DOM. In many cases, we need the ability to act when particular events occur. Angular offers *component lifecycle hooks* to allow us observe these events and respond them.

In the refactored version of **DeleteSiteComponent**, we use the **OnChanges** lifecycle hook. This hook provides a single method, **ngOnChanges** that is invoked whenever Angular detects the changes to input properties of the component. Listing 3-10 shows how **DeleteSiteComponent** utilizes this lifecycle hook.

Listing 3-10: edit-site.component.ts (Exercise-03-06)

```
import {Component, Input, Output, EventEmitter} from '@angular/core';
import {OnChanges} from '@angular/core';
import {SiteManagementService} from './site-management.service'

@Component({
  selector: 'delete-site-view',
  templateUrl: 'app/delete-site.template.html'
})
export class DeleteSiteComponent implements OnChanges{
  @Input() siteId: number;
  @Output() onClosed = new EventEmitter();
  siteName: string;

  constructor(private siteService: SiteManagementService) {
  }

  ngOnChanges() {
    this.siteName = this.siteService
      .getSiteById(this.siteId).name;
  }

  delete() {
    this.siteService.deleteSite(this.siteId);
    this.onClosed.emit(null);
  }

  cancel() {
    this.onClosed.emit(null);
  }
}
```

When there is time to invoke a lifecycle hook, Angular just checks whether to component instance has a method that matches the name of the lifecycle hook operation, and calls it. When any input property of **DeleteSiteComponent** changes its value, Angular invokes **ngOnChanges()**—and this method initializes **siteName** with the **getSiteById()** service call.

Though it had been enough to add **ngOnChanges()** method to a component, Listing 3-10 did it the suggested way. **DeleteSiteComponent** implements the **OnChanges** interface that defines the **ngOnChanges** method:

```
export class DeleteSiteComponent implements OnChanges {
  // ...
  ngOnChanges() {
    // ...
  }
  // ...
}
```

Although Angular works without the interface implementation, I recommend you to apply this pattern, too. First, the **implements OnChanges** declaration at the end of the class definition is an explicit sign for the readers of the code that this class handles this lifecycle hook. Second, most code editors benefit from the strong typing support provided by TypeScript. For example, in Visual Studio Code, if you forgot about adding the **ngOnChanges()** method to the class definition, the editor would warn you about it (Figure 3-9).

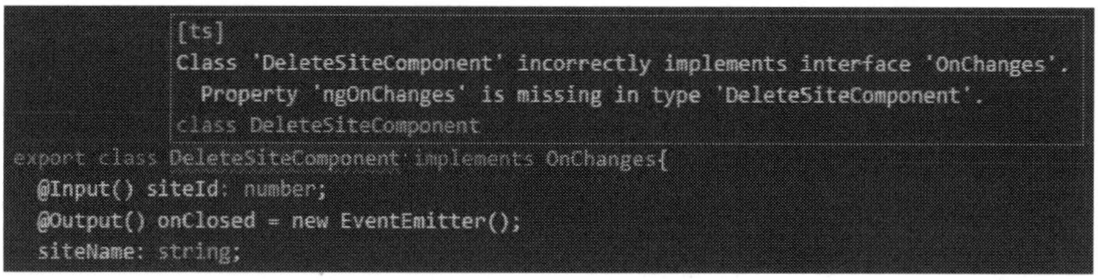

```
[ts]
Class 'DeleteSiteComponent' incorrectly implements interface 'OnChanges'.
  Property 'ngOnChanges' is missing in type 'DeleteSiteComponent'.
class DeleteSiteComponent
export class DeleteSiteComponent implements OnChanges{
  @Input() siteId: number;
  @Output() onClosed = new EventEmitter();
  siteName: string;
```

Figure 3-9: Strong typing helps the editor to observe issues

With these changes, the dive site maintenance application is refactored to use a service that encapsulates all data management responsibilities. This architecture would help a lot when you intend to move these data services to the backend.

Module Injectors

Before going on, let's stop for a moment a discuss how the framework injected **SiteManagementService** into components. So far, we used the **providers** metadata property of **AppComponent** to tell Angular we want to pass the same singleton instance of **SiteManagementService** to all child view—including list, add, edit, and delete:

```
import {Component} from '@angular/core';
import {SiteManagementService} from './site-management.service'

@Component({
  selector: 'yw-app',
```

```
    templateUrl: 'app/app.template.html',
    providers: [SiteManagementService]
})
export class AppComponent {
  // ...
}
```

Besides components, Angular modules (**NgModules**) have their injectors. When the dependency injection mechanism tries to resolve a requested dependency, it checks the module injector, too. Instead of declaring the provider in **AppComponent**, we could move it to the **AppModule**. From now on, I will follow this modified pattern (Listing 3-11, Listing 3-12).

Listing 3-11: app.component.ts (Exercise-03-07)

```
import {Component} from '@angular/core';

@Component({
  selector: 'yw-app',
  templateUrl: 'app/app.template.html'
})
export class AppComponent {
  // ...
  }
}
```

Listing 3-12: app.module.ts (Exercise-03-07)

```
import {NgModule} from '@angular/core';
import {BrowserModule} from '@angular/platform-browser';

import {AppComponent} from './app.component';
import {SiteListComponent} from './site-list.component';
import {AddSiteComponent} from './add-site.component';
import {EditSiteComponent} from './edit-site.component';
import {DeleteSiteComponent} from './delete-site.component';
import {SiteManagementService} from './site-management.service';

@NgModule({
  imports: [BrowserModule],
  declarations: [
    AppComponent,
    SiteListComponent,
    AddSiteComponent,
    EditSiteComponent,
    DeleteSiteComponent
  ],
  providers: [SiteManagementService],
  bootstrap: [AppComponent]
})
export class AppModule { }
```

Using Attribute Directives

You already learned about the pivotal concept of Angular, the *component*. In the previous section, you got acquainted with the fundamental benefits of services. In this section, you will get to know another Angular idea, *attribute directives*.

In *Chapter 2*, you met with several directives, including `ngIf`, `ngFor`, `ngSwitch`, `ngStyle`, and `ngClass`. So far, I did not mention that components are directives, too. Angular has three kinds of directives:

Components. These are directives that have templates, and so manage their dedicated patch of UI. When we build Angular applications, most of the time we design and create components.

Structural directives. These change the layout of the DOM either by adding or removing DOM elements. `ngIf`, `ngFor`, and `ngSwitch` are great examples. Structural directives do not have templates. We can create our custom structural directives, but we seldom need to do.

Attribute directives. Attribute directives can be attached to DOM elements, and although they do not have templates, they can change the appearance or behavior of elements to which they are annexed.

Creating a Simple Attribute Directive

In this section, we are going to create a new attribute directive that can be used to attach actions to a DOM element. We are going to use this directive to highlight the DOM element as the mouse pointer moves over it to indicate that it is actionable, and execute the specified action when the user clicks it.

To create this attribute directive, follow these steps:

1. Copy the entire **Exercise-03-07** folder into a new folder (**Exercise-03-08),** and work with the content of the new folder in the subsequent steps.

2. Create a new file, **actionable.directives.ts**, in the **app** folder, and type this code into it:

```
import { Directive, ElementRef, HostListener } from '@angular/core';

@Directive({
  selector: '[ywActionable]'
})
export class ActionableDirective {

  constructor(private element: ElementRef) {
  }

  @HostListener('mouseenter') onMouseEnter() {
    this.setAppearance('#aaaaaa', 'pointer');
  }
```

```
  @HostListener('mouseleave') onMouseLeave() {
    this.setAppearance(null, null);
  }

  setAppearance(color: string, cursor: string) {
    let style = this.element.nativeElement.style;
    style.backgroundColor = color;
    style.cursor = cursor;
  }
}
```

3. Add the `ywActionable` attribute to `SiteListComponent's` template (`site-list.template.html`):

```
...
<h2>List of Dive Sites</h2>
<div class="row" *ngFor="let site of sites; let e=even; let f=first; let
l=last"
  ywActionable
  [ngClass]="{ evenRow: e }"
  [class.topRow]="f"
  [class.bottomRow]="l">
<!-- ... -->
</div>
```

4. Open `app.module.ts`, and add the new `ActionableDirective` class to the module declarations:

```
import {NgModule} from '@angular/core';
import {BrowserModule} from '@angular/platform-browser';

import {AppComponent} from './app.component';
import {SiteListComponent} from './site-list.component';
import {AddSiteComponent} from './add-site.component';
import {EditSiteComponent} from './edit-site.component';
import {DeleteSiteComponent} from './delete-site.component';
import {SiteManagementService} from './site-management.service';
import {ActionableDirective} from './actionable.directive';

@NgModule({
  imports: [BrowserModule],
  declarations: [
    AppComponent,
    SiteListComponent,
    AddSiteComponent,
    EditSiteComponent,
    DeleteSiteComponent,
    ActionableDirective
  ],
  providers: [SiteManagementService],
  bootstrap: [AppComponent]
})
```

```
export class AppModule { }
```

5. Run the app. As you move the mouse over the list of dive sites, you can see that the background gets a darker shade, and the cursor changes to a pointing hand (Figure 3-10)—and this is what we expect the **ywActionable** directive to do.

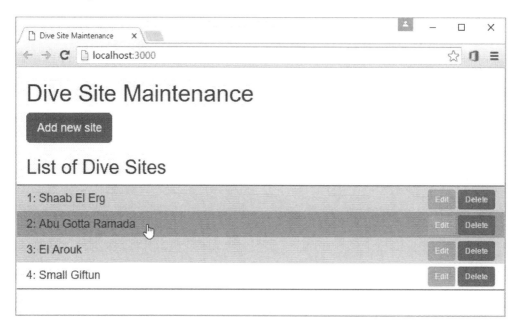

Figure 3-10: The ywActionable attribute directive in action

Let's see, how this directive works!

In step 2, you created a new class decorated with the **@Directive** annotation:

```
import { Directive, ElementRef, HostListener } from '@angular/core';

@Directive({
  selector: '[ywActionable]'
})
export class ActionableDirective {

  constructor(private element: ElementRef) {
  }

  // ...
}
```

The **@Directive** annotation adds the **selector** metadata property to the class. The selector's value is "[ywActionable]" so this directive will match with any HTML element that has a **ywActionable** attribute.

> **NOTE**: *In the selector of the attribute directive, I used the "yw" (Younderwater) prefix to mitigate the risk that it collides with potential third-party directive names. Angular marks its directives with the "ng" prefix.*

The constructor accepts an **ElementRef** instance that represents the DOM elements to which the **ywActionable** attribute is assigned. **ActionableDirective** provides its behavior with these methods:

```
@HostListener('mouseenter') onMouseEnter() {
  this.setAppearance('#aaaaaa', 'pointer');
}

@HostListener('mouseleave') onMouseLeave() {
  this.setAppearance(null, null);
}

setAppearance(color: string, cursor: string) {
  let style = this.element.nativeElement.style;
  style.backgroundColor = color;
  style.cursor = cursor;
}
```

When the framework meets a method marked with the **@HostListener()** decorator, it invokes the decorated method when the host element of the attribute directive emits the specified event. Here, the directive attaches its methods to the **mouseenter** and **mouseleave** events of the host tag. These invoke **setAppearance()** with a color and a cursor value. The **ElementRef** instance the directive got in its constructor is a wrapper around the real DOM element that can be accessed via the **nativeElement** property. In **setAppearance()**, the code utilizes **nativeElement** to set the style properties of the host element.

In step 4, you added the declaration of **ActionableDirective** to module declaration so that Angular can bound it to the **ywActionable** attribute. Without this step, the framework would not be able to find the host for **ywActionable**.

Improving the Attribute Directive

At the moment, **ActionableDirective** does not provide any action. Let's enhance it with input property binding that allows setting the color of the background, and an event binding that provides an action when the user clicks the host element.

The solution is very straightforward; you can examine Listing 3-13.

Listing 3-13: actionable.directive.ts (Exercise-03-09)

```
import { Directive, ElementRef, HostListener } from '@angular/core';
import { Input, Output, EventEmitter } from '@angular/core';

@Directive({
  selector: '[ywActionable]'
})
export class ActionableDirective {
  @Input('ywActionable') backgroundColor: string;
  @Output() onAction = new EventEmitter();

  constructor(private element: ElementRef) {
  }

  @HostListener('mouseenter') onMouseEnter() {
    this.setAppearance(this.backgroundColor || 'green', 'pointer');
  }

  @HostListener('mouseleave') onMouseLeave() {
    this.setAppearance(null, null);
  }

  @HostListener('click') onClick() {
    this.onAction.emit(null);
  }

  setAppearance(color: string, cursor: string) {
    let style = this.element.nativeElement.style;
    style.backgroundColor = color;
    style.cursor = cursor;
  }
}
```

As you see from this listing, you can use input property binding and output event emitting with attribute directives the same way you do it with components. Hey, it is logical—a component is a directive!

Nonetheless, the code demonstrates a new feature. When you add the **@Input()** or **@Output()** decorator to a class member, you can provide an optional name for the attribute that represents the corresponding property or event. This is how we map the value of the **ywActionable** attribute's value to the **backgroundColor** field:

```
@Input('ywActionable') backgroundColor: string;
```

We could apply a separate attribute name for declaring the background color, but the markup is more concise if we do not. The template of **SiteListComponent** now declares the **ywActionable** attribute this way:

```
<div class="row" *ngFor="let site of sites; let e=even; let f=first; let
l=last"
  ywActionable="#6666ee" (onAction)="edit(site.id)"
  [ngClass]="{ evenRow: e }"
  [class.topRow]="f"
  [class.bottomRow]="l">
```

When you run the app, you can click the items directly in the dive site list—not only the Edit button—and you get to the edit view.

Routing

A significant amount of the code in the dive site maintenance application deals with navigation from one view to another. **AppComponent** receives event notifications from other views and in response, it navigates to the next view. Components apply input and output properties and emit events to keep the navigation mechanism working.

The sample app has only four views. If we were dealing with a dozen of them, the technique we have been using so far would make the app unnecessarily complicated. We would add many code lines just for the sake of bolstering the navigation.

Angular offers another approach based on its *component router*. Instead of wiring up components through properties, this model parses browser URLs and interprets them as navigation instructions. In this section, you will learn the fundamentals of Angular routing—through the refactoring of the dive site maintenance app.

This kind of refactoring is a notable change in the architecture—fortunately, it makes the code more concise—, thus we cannot do it view-by-view, it is an all-or-nothing modification.

Using Navigation URLs

The component router leverages in-app URL paths that look like server URLs. We can teach the component router to understand these in-app URLs and navigate to the appropriate component that can handle a particular operation. Provided the app's base URL is **http://localhost:3000**— as in the samples of the book's code download—, we want the router to interpret URLs this way:

In response to **http://localhost:300/site**, the app should navigate to **SiteListComponent**. Similarly, **http://localhost:3000/add** should trigger the navigation to **AddSiteComponent**. When we intend to edit a dive site or remove one, we need to pass the id of the corresponding site as the part of the URL. Thus we will use the **http://localhost:3000/edit/id** and **http://localhost:3000/delete/id** URLs, respectively. Of course, in place of *id*, we pass a particular identifier, such as **3** or **7**.

Preparing the Component Router

To manage the navigation URLs correctly, we need to prepare the component router. First of all, we have to add a **<base>** tag to our **index.html** file (Listing 3-14); otherwise, the router will not work.

Listing 3-14: index.html (Exercise-03-10)

```
<!DOCTYPE html>
<html>
<head>
  <base href="/">
  <title>Dive Site Maintenance</title>
  <link href="/node_modules/bootstrap/dist/css/bootstrap.min.css"
    rel="stylesheet" />
  <script src="/node_modules/core-js/client/shim.min.js"></script>
  <script src="/node_modules/zone.js/dist/zone.js"></script>
  <script src="/node_modules/reflect-metadata/Reflect.js"></script>
  <script src="/node_modules/systemjs/dist/system.src.js"></script>

  <script src="systemjs.config.js"></script>
  <script>
    System.import('app').catch(function(err){ console.error(err); });
  </script>
</head>
<body>
  <yw-app>Loading...</yw-app>
</body>
</html>
```

Second, we have to tell the component router how to parse our URLs. Just like as physical network routers are configured with a routing table, we need to create such a table for the component router. It is a good practice to place these routing definitions into a separate file. In the refactored sample, I used the **app.routes.ts** file, as shown in Listing 3-15.

Listing 3-15: app.routes.ts (Exercise-03-10)

```
import {Routes, RouterModule} from '@angular/router';
import {SiteListComponent} from './site-list.component';
import {AddSiteComponent} from './add-site.component';
import {EditSiteComponent} from './edit-site.component';
import {DeleteSiteComponent} from './delete-site.component';

const routes: Routes = [
  { path: 'list', component: SiteListComponent },
  { path: 'add', component: AddSiteComponent },
  { path: 'edit/:id', component: EditSiteComponent },
  { path: 'delete/:id', component: DeleteSiteComponent },
```

```
    { path: '', redirectTo: 'list', pathMatch: 'full'}
];

export const routingModule = RouterModule.forRoot(routes);
```

The **Routes** class can be used to define the routing table for the component router. Each entry is a **Route** instance that describes how an individual URL should be interpreted and what action the router should take. The first four definition uses the **path** and **component** properties to declare that a certain **path**—relative to the base URL, namely the URL specified in **<base>**—should display the view of a particular **component**. The third and fourth entries define the paths as "edit/:id" and "delete/:id", respectively. The ":id" parts represent path parameters; we use them to pass the id of a particular site to the destination components.

The last URL is to redirect the **http://localhost:3000** URL automatically to **http://localhost:3000/list**. Because the empty path matches with any URL, setting **pathMatch** to "full" ensures that the router will apply the redirect if and only if navigating to base URL.

We export the **routingModule** variable the value of which is retrieved by **RouterModule.forRoot()**. As its name suggests, **RouterModule** is an Angular module that implements the component router and its accessory types. The **forRoot()** method dynamically creates a module that is configured to support routing according to the routing table passed as the input argument. The routing consumes several services. These are prepared by **forRoot()**, too.

> *TypeScript: The **const** keyword declares a variable that can be assigned only at the time of its declaration; it prevents the re-assignment of the variable.*

Now, that we have defined the routes to be used with the app, we need to tell Angular to apply them. We do it by importing the exported **routingModule** in the application's root module declaration (Listing 3-16).

Listing 3-16: app.modules.ts (Exercise-03-10)

```
import {NgModule} from '@angular/core';
import {BrowserModule} from '@angular/platform-browser';
// ...
import {routingModule} from './app.routes';

@NgModule({
  imports: [
    BrowserModule,
    routingModule
  ],
  declarations: [
    AppComponent,
    SiteListComponent,
    AddSiteComponent,
    EditSiteComponent,
```

```
    DeleteSiteComponent,
    ActionableDirective
  ],
  providers: [SiteManagementService],
  bootstrap: [AppComponent]
})
export class AppModule { }
```

Now, you can run the application. However, when you try to navigate to a URL, let's say to "/add", the browser drops the page back to the list view with an error message (Figure 3-11).

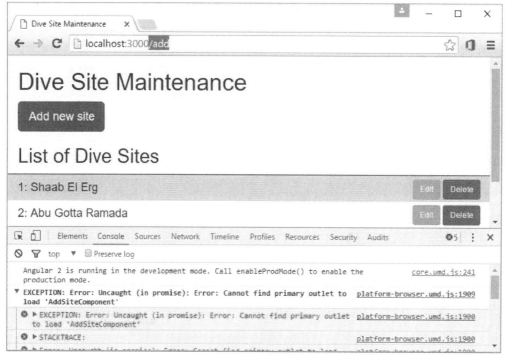

Figure 3-11: Navigation does not work yet

> **NOTE**: You do not need to navigate to "/add", the error messages come out when you start the app. Nonetheless, you do not see them unless you display console messages.

To display the component the app is routed to, we need to add a router outlet.

Adding a Router Outlet

Now it is time to refactor the existing components to leverage the benefits of the component router. It requires changing all component classes and their templates, too.

First, let's remove the error message displayed in Figure 3-11!

The error message tells us that we need to specify where the component the current page is routed to should display its view. To do that, we need to add a **`<router-outlet>`** element to **`AppComponent`**'s template, as Angular shows the current components view within the bounds of **`<router-outlet>`**.

So far, **`AppComponent`** implemented the logic—with the help of the **`ngSwitch`** directive—that displayed the current view (list, add, edit, or delete). Now, the component router undertakes this role, so **`AppComponent`** and its template become very lean (Listing 3-17, Listing 3-18).

Listing 3-17: app.component.ts (Exercise-03-11)

```
import {Component} from '@angular/core';

@Component({
  selector: 'yw-app',
  templateUrl: 'app/app.template.html',
})
export class AppComponent {
}
```

Listing 3-18: app.template.html (Exercise-03-11)

```
<div class="container-fluid">
  <h1>Dive Site Maintenance</h1>
  <router-outlet></router-outlet>
</div>
```

When you run the app, you can navigate to any of the predefined URLs, and the app shows the appropriate component. Of course, they do not work yet properly, but at least their views are displayed on the page. Figure 3-12 the result of visiting the "/edit/1" URL.

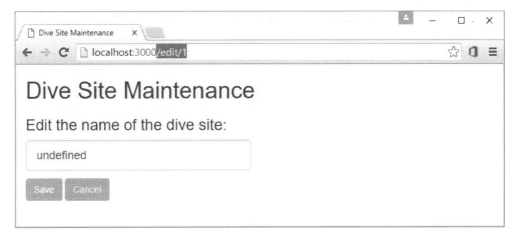

Figure 3-12: The edit view is accessed with a URL

The component router accepts only the URLs it understands. If you try to use URLs that do not match with the routing table, you get error messages (Figure 3-13).

Figure 3-13: Navigating to "/blabla" results in error

NOTE: *If you try to navigate to "/edit", you will receive the same error message because "/edit" does not match with any rule. You should use a URL that matches the "/edit/:id" pattern.*

Navigating with Links

The component router keeps the browser's location URL and the current path synchronized. When the user changes the browser's location URL, the router selects the appropriate component and displays the corresponding view. When the app requests navigation to a particular path—provided it is valid and understood by the component router—, besides presenting the destination component's view, the browser's location—and the navigation history—is updated.

The refactored `SiteListComponent` is an excellent example for demonstrating how you can navigate through links in the component template—and programmatically within the component's class definition.

`SiteListComponent` should be able to navigate to the add, edit, and delete views, respectively. In the previous section, we added the `ywActionable` directive that allows navigating to the edit view. `ywActionable` executes an action, so we provide a programmatic way to jump to the edit view, too.

Listing 3-19 shows the changes in the template of **SiteListComponent**. Observe, this code uses **<a>** elements instead of **<button>**—due to the Bootstrap library, **<a>** and **<button>** have the same appearance—, and leverages a new routing directive, **routerLink**.

Listing 3-19: site-list.template.html (Exercise-03-12)

```
<div class="row">
  <div class="col-sm-12">
    <a class="btn btn-primary btn-lg"
      routerLink="/add">
      Add new site
    </a>
  </div>
</div>
<h2>List of Dive Sites</h2>
<div class="row" *ngFor="let site of sites; let e=even; let f=first; let
l=last"
  ywActionable="#aaaaaa" (onAction)="edit(site.id)"
  [ngClass]="{ evenRow: e }"
  [class.topRow]="f"
  [class.bottomRow]="l">
  <div class="col-sm-8">
    <h4>{{site.id}}: {{site.name}}</h4>
  </div>
  <div class="col-sm-4" style="margin-top: 5px;">
    <div class="pull-right">
      <a class="btn btn-warning btn-sm"
        [routerLink]="['/edit', site.id]">
        Edit
      </a>
      <a class="btn btn-danger btn-sm"
        [routerLink]="['/delete', site.id]">
        Delete
      </a>
    </div>
  </div>
</div>
```

To navigate to a configured root, we can use the **routerLink** directive. Its template expression should be a string or an array; the directive will resolve it as a URL. If a string is provided, it represents the route, such as in **routerLink="/add"**. If we use an array, the first element is the route, and the subsequent items are route parameters. In the code above, provided **site.id** is 5, the **[routerLink]="['/edit', site.id]"** results in **/edit/5**.

We can access the router functionality from code (Listing 3-20).

Listing 3-20: site-list.component.ts (Exercise-03-12)

```
import {Component} from '@angular/core';
import {Router} from '@angular/router';

import {DiveSite} from './dive-site';
import {SiteManagementService} from './site-management.service'

@Component({
  selector: 'site-list-view',
  templateUrl: 'app/site-list.template.html',
  styleUrls: ['app/site-list.styles.css']
})
export class SiteListComponent {
  sites: DiveSite[];

  constructor(
    private siteService: SiteManagementService,
    private router: Router
  ) {
    this.sites = siteService.getAllSites();
  }

  edit(siteId: number) {
    this.router.navigate(['/edit', siteId]);
  }
}
```

In the edit method of **SiteListComponent**, we navigate to the edit view programmatically, and thus we need to access the component router. Because the component router is a service, we can access it through constructor injection as shown in the listing. The edit method invokes the **navigate()** method of the router with the route described as an array—just as in the **routerLink** template expression.

> *NOTE: I moved the style definitions into a separate stylesheet file,* **site-list.styles.css**.

Now, when you start the app, you can navigate from the list view to add, edit, and delete, however, none of them works yet as expectable.

Let's complete the refactoring by moving all the other views to the new navigation model!

Refactoring the Add View

We can use the approach we have just learned to modify the add view, as the changes of **AddSiteComponent** and its template are shown in Listing 3-21 and Listing 3-22.

Listing 3-21: add-site.component.ts (Exercise-03-13)

```
import {Component, EventEmitter} from '@angular/core';
import {Router} from '@angular/router';

import {SiteManagementService} from './site-management.service'

@Component({
  selector: 'add-site-view',
  templateUrl: 'app/add-site.template.html'
})
export class AddSiteComponent {
  siteName: string;

  constructor(
    private siteService: SiteManagementService,
    private router: Router) { }

  add() {
    this.siteService.addSite({id: 0, name:this.siteName});
    this.router.navigate(['/list']);
  }
}
```

Listing 3-22: add-site.template.html (Exercise-03-13)

```
<h3>Specify the name of the dive site:</h3>
<div class="row">
  <div class="col-sm-6">
    <input #siteNameBox class="form-control input-lg" type="text"
      placeholder="site name"
      (keyup)="siteName=siteNameBox.value"
      (keyup.enter)="added()" />
  </div>
</div>
<div class="row" style="margin-top: 12px;">
  <div class="col-sm-6">
    <button class="btn btn-success btn"
      (click)="add()"
      [disabled]="!siteName">
      Add
    </button>
    <a class="btn btn-warning btn"
      routerLink="/list">
      Cancel
    </a>
  </div>
</div>
```

Refactoring the Edit and Delete Views

Thanks to **routerLink**, we need to apply only slight changes to the template of
EditSiteComponent (Listing 3-23).

Listing 3-23: edit-site.template.html (Exercise-03-13)

```
<h3>Edit the name of the dive site:</h3>
<div class="row">
  <div class="col-sm-6">
    <input #siteNameBox class="form-control input-lg" type="text"
      [value]="siteName"
      placeholder="site name"
      (keyup)="siteName=siteNameBox.value"
      (keyup.enter)="save()" />
  </div>
</div>
<div class="row" style="margin-top: 12px;">
  <div class="col-sm-6">
    <button class="btn btn-success btn"
      (click)="save()"
      [disabled]="!siteName">
      Save
    </button>
    <a class="btn btn-warning btn"
      routerLink="/list">
      Cancel
    </a>
  </div>
</div>
```

When we navigate to the edit view, we need to access the id of the site to allow editing the site
name. Of course, we would not like to analyze the URL of the route. We do not have to: the
component router does it for us. Listing 3-24 shows how we can obtain the current value of the
site id route parameter.

Listing 3-24: edit-site.component.ts (Exercise-03-13)

```
import {Component} from '@angular/core';
import {Router, ActivatedRoute} from '@angular/router';

import {SiteManagementService} from './site-management.service'

@Component({
  selector: 'edit-site-view',
  templateUrl: 'app/edit-site.template.html'
})
export class EditSiteComponent {
  siteId: number;
  siteName: string;
```

```
    private parSub: any;

    constructor(
      private siteService: SiteManagementService,
      private route: ActivatedRoute,
      private router: Router
    ) {
      this.siteId = this.route.snapshot.params['id'];
      this.siteName = this.siteService
        .getSiteById(this.siteId).name;
    }

    save() {
      this.siteService.saveSite({id: this.siteId, name:this.siteName});
      this.router.navigate(['/list']);
    }
}
```

In the constructor, besides the **Router**, we inject the **ActivatedRoute** service. The latter one provides access to route parameters, and we obtain the site id this way:

```
this.siteId = this.route.snapshot.params['id'];
```

The **snapshot** property of **ActivateRoute** allows accessing the *initial* parameters of the route, and we can access them by their name.

> *NOTE: The component router can re-use the components—it can use the same component instance as the target for multiple navigations—, and in this case instead of using the snapshot of initial parameters, we can subscribe to route parameter changes. In this chapter, we do not discuss this advanced scenario, but later in the book, you will learn about it.*

The delete view has the same structure as the edit view, but instead of saving the site information, it removes it. You can find the refactored version in the **delete-site.component.ts** and **delete-site.template.html** files within the **Exercise-03-13** folder.

Summary

Angular has the concept of services. A *service* is a class with a clear responsibility, a narrow set of operations proffered to other components. Many operations of the framework can be accessed through services and, of course, we can create our custom services to improve the modularity of apps.

Angular has three types of directives. *Components* are one of them, and we can implement *structural directives* and *attribute directives*, too. In this chapter, you learned that attribute directives can be attached to DOM elements to modify the particular elements' appearance and behavior.

Angular modules help organize an application into cohesive blocks of functionality. The framework itself has many modules, and many third-party libraries are also available as modules.

Most single page apps need to navigate among views according to the app's workflow. The Angular component router provides a straightforward way that leverages in-app URLs. With the related set of routing services, you can quickly move from one view to another one either declaratively or programmatically.

In the next chapter, you will learn how to access backend services through HTTP.

Chapter 4: Accessing Backend Services

WHAT YOU WILL LEARN IN THIS CHAPTER

Understanding the problems raised by communicating with backend services

Getting acquainted with the significance of asynchronous program execution

Learning the basics of JavaScript promises

Using Angular's `Http` service

Building a simple application that uses a REST-like backend API with authorization

Most web apps use backend services that store and retrieve data. Without accessing the data or related operations that can manipulate the information, no real applications can work.

You can either underestimate or overestimate the complexity of accessing data stored in the backend. If you think this activity as simple as invoking a method, you certainly underestimate the work to be done. Similarly, if you start thinking about the details of communication protocols, and envisioning the issues that should be solved to make your backend access work, you probably overestimate the labor you need to invest.

Similarly to jQuery, Angular has its toolset to make it straightforward to access backend services through the HTTP or HTTPS protocols—while it still allows you to configure and customize the way your requests are sent to the server.

In this chapter, you will learn and understand how to communicate with backend services—the Angular way.

Refactoring the Dive Log Application

In *Chapter 1, Moving to Angular*, you already got acquainted with the basic concepts of Angular through an app that displayed entries from a diver's logbook. Here, you will change this app so that it will obtain the data from a real server running somewhere in the cloud. Instead of jumping right away to the final version of this sample, you will migrate the logbook app in a few steps, each of which allows you to learn a few basic things about managing web requests. As the initial step, you will take a look at the refactored form of `Exercise-01-04`.

From the architecture point of view, it is always a good idea to outsource the responsibility of calling backend APIs into a separate component. As you learned in *Chapter 3, Using Services*, Angular services are excellent candidates to encapsulate such functionality.

The **Exercise-04-01** folder of this chapter's source code download contains the refactored version of the logbook app. Now, it has a Refresh button. When you click it (Figure 4-1), the page displays a progress ring to mimic the backend operation—it waits for one second—, and then it shows up three dive log entries (Figure 4-2).

> **NOTE**: *This sample—and all subsequent exercises—uses the* application root module *pattern you learned in the previous chapter.*

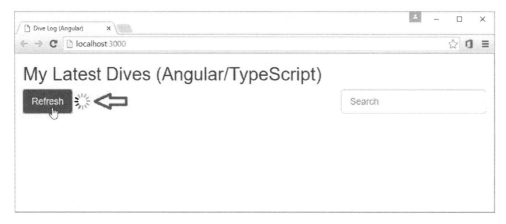

Figure 4-1: The backend operation is mimicked with a progress ring

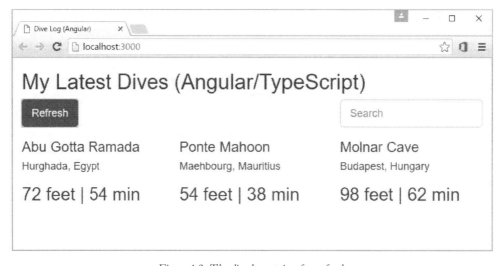

Figure 4-2: The dive log entries after refresh

This sample uses a tiny Angular service to separate the backend access from the app (Listing 4-1). The **getDives()** operation retrieves the pre-defines list of dive log entries.

Listing 4-1: dive-log-api.service.ts (Exercise-04-01)

```
import {Injectable} from '@angular/core';
import {DiveLogEntry} from './dive-log-entry';

@Injectable()
export class DiveLogApi {
  getDives() {
    return DiveLogEntry.StockDives;
  }
}
```

Clicking the Refresh button invokes the **refreshDives()** method in **DiveLogComponent**, as shown in Listing 4-2. This method leverages the standard JavaScript **setTimeout()** operation to delay the loading of log records and utilizes the **loading** flag to control the visibility of the progress ring.

Listing 4-2: dive-log.component.ts (Exercise-04-01)

```
import {Component} from '@angular/core';
import {ContentFilterPipe} from './content-filter.pipe'
import {DiveLogEntry} from './dive-log-entry';
import {DiveLogApi} from './dive-log-api.service'

@Component({
  selector: 'divelog',
  templateUrl: 'app/dive-log.template.html'
})
export class DiveLogComponent {
  loading = false;
  dives: DiveLogEntry[];

  constructor(private api: DiveLogApi) {
  }

  refreshDives() {
    this.loading = true;
    this.dives = [];
    setTimeout(() => {
      this.dives = this.api.getDives();
      this.loading = false;
    }, 1000);
  }
}
```

> **NOTE:** In the listing, I highlighted the lines that are required to access the **DiveLogApi** within the component. Observe, that we do not need to add **providers** metadata property to the component, for the **DiveLogApi** service is added to the providers of the application root module in **app.module.ts**. Similarly, we do not need to add **pipes** metadata to the component.

This sample applied the asynchronous programming pattern: the **refreshDives()** method returns immediately after the **setTimeout()** call, although the highlighted code runs only a second later—when the preset timeout expires.

The asynchronous programming pattern is very attractive. One of the main reason for its attractiveness is that it allows keeping the UI responsive while doing some work in the background. When you are about to access backend services, asynchrony is especially useful: while waiting for the server to retrieve the results, the browser keeps the UI responsive.

TypeScript: Lambda functions

The **this** keyword is JavaScript needs attention because it can be a source of issues when we use it within anonymous functions. If we had used the traditional JavaScript syntax in **refreshDives()**, our application would break, and raise an error message (Figure 4-3):

```
refreshDives () {
    this.loading = true;
    this.dives = [];
    setTimeout(function() {
        this.dives = this.api.getDives();
        this.loading = false;
    }, 1000);
}
```

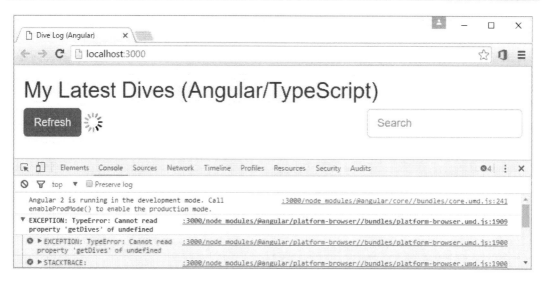

Figure 4-3: Error message when using the traditional JavaScript syntax

With this syntax, we should have written this code to ensure proper operation:

```
refreshDives() {
  this.loading = true;
  this.dives = [];
  var component = this;
  setTimeout(function() {
    component.dives = component.api.getDives();
    component.loading = false;
  }, 1000);
}
```

To make things simpler, in TypeScript we can use the lambda syntax, and then the compiler will handle the **this** keyword according to our intention:

```
refreshDives() {
  this.loading = true;
  this.dives = [];
  setTimeout(() => {
    this.dives = this.api.getDives();
    this.loading = false;
  }, 1000);
}
```

> **HINT**: *Think it over twice to convert between the traditional JavaScript syntax and TypeScript lambda function notation (and vice versa), especially when you see **this** being used within the function body! Whenever you can, use lambdas.*

Working with Promises

This book is not about to teach you asynchronous programming. Without diving deeply into the details, asynchronous implementation starts to become more involved when your code execution flow gets complicated, includes asynchronous blocks nested into other asynchronous blocks. Exception handling is another challenging issue.

There are many useful programming patterns to handle asynchrony. In the JavaScript world, the *promise pattern* is the most popular and wide-spread. A *promise* represents the result of a potentially long running and not necessarily complete operation. Instead of blocking and waiting for the long-running computation to complete, the pattern returns an object which represents the promised result.

At any moment in time, a promise can be in one of three states: *unfulfilled, resolved* or *rejected*. To have a better understanding what these mean, let's interpret them in the case of a backend service call.

When you start a backend call with a promise, the promise gets into the *unfulfilled* state. The client sends a request to the server and waits for the server-side response. If serving the request takes a long time, the user may close the page (or the browser), and the promise never leaves its unfulfilled state.

When the server responds with a status that indicates a successful operation, the promise gets into *resolved* state. If the server signs an error status, the promise moves into *rejected* state.

> *NOTE: The unfulfilled state means an operation still in progress or an event that has not occurred yet. The resolved and rejected states represent the outcome of an operation or an event.*

There are several ways you can use promises, the most frequently used form is this:

```
promiseOperation()
  .then(resolvedHandler, rejectedHandler)
```

Here, **promiseOperation()** is a function that returns a promise object. You can invoke the **then()** method on that object. **then()** accepts two functions as arguments. The first argument (**resolvedHandler**) defines the handler for the event when the promise is resolved, while the second—optional—argument (**rejectedHandler**) describes the event for the rejected promise.

Both **resolvedHandler** and **rejectedHandler** may use arguments that are specific for **promiseOperation()**.

> *NOTE: A promise is not just another variation of an event handler. While event handlers respond all occurrences of events, a promise responds only once. Its outcome activates either the resolved or the rejected handler—or stays in the unfulfilled state. When any of the outcomes is fulfilled, the promise reaches the end of its life cycle.*

Refactoring the App to Use Promises

I guess, this short overview of promises may be a bit dry for you. It is better to take a look at a concrete example so that you get acquainted with this pattern through a refactored version of the logbook app.

In the **Exercise-04-02** folder, you can find the modified app. Now, it uses the generic form of **Promise**, which is the part of the standard TypeScript library—and uses the **Promise** JavaScript object (Listing 4-3).

Listing 4-3: dive-log-api.service.ts (Exercise-04-02)

```
import {Injectable} from '@angular/core';
import {DiveLogEntry} from './dive-log-entry';
```

```
@Injectable()
export class DiveLogApi {
  getDives() {
    return new Promise<DiveLogEntry[]>((resolve, reject) => {
      setTimeout(() => {
        resolve(DiveLogEntry.StockDives);
      }, 1000);
    })
  }
}
```

The **getDives()** function returns a promise that is resolved one second after the promise is retrieved. Because we use a generic promise with the type parameter of **DiveLogEntry[]**, the **resolve()** function is expected to have an argument of the very same type—the TypeScript compiler checks it.

Now, the **DiveLogApi** returns a promise, so the **refreshDives()** function should be modified accordingly, as shown in Listing 4-4.

Listing 4-4: dive-log.component.ts (Exercise-04-02)

```
...
export class DiveLogComponent {
  loading = false;
  dives: DiveLogEntry[];

  constructor(private api: DiveLogApi) {
  }

  refreshDives() {
    this.loading = true;
    this.dives = [];

    this.api.getDives()
      .then(data => {
        this.dives = data
        this.loading = false;
      });
  }
}
```

The **refreshDives()** method returns very quickly, almost in the same instance when the user clicks the Refresh button. Because **getDives()** retrieves a promise, the function passed to **then()** is executed when the promise is resolved—one second after the click. The resolved promise provides an array of **DiveSite** objects. This array is received in **data** and then assigned to **dives**. Angular detects the change and updates the UI.

Rejecting the Promise

The **Exercise-04-03** folder holds an updated sample to demonstrate the rejection of a promise. In this version, the **DiveLogApi** component raises an error after each third **getDives()** call (Figure 4-4).

Figure 4-4: *The result of a rejected promise*

The modified version of **DiveLogApi** invokes the **reject()** method (Listing 4-5).

Listing 4-5: *dive-log-api.service.ts (Exercise-04-03)*

```
import {Injectable} from '@angular/core';
import {DiveLogEntry} from './dive-log-entry';

@Injectable()
export class DiveLogApi {
  static counter = 0;
  getDives() {
    return new Promise<DiveLogEntry[]>((resolve, reject) => {
      DiveLogApi.counter++;
      setTimeout(() => {
        if (DiveLogApi.counter % 3 == 0) {
          reject(`Error: Call counter is ${DiveLogApi.counter}`);
        } else {
          resolve(DiveLogEntry.StockDives);
        }
      }, 1000);
    })
  }
}
```

DiveLogComponent should be prepared for managing the rejection. As Listing 4-6 shows, it obtains the error message raised by **getDives()**.

Listing 4-6: dive-log.component.ts (Exercise-04-03)

```
...
export class DiveLogComponent {
  loading = false;
  errorMessage = null;
  dives: DiveLogEntry[];

  constructor(private api: DiveLogApi) {
  }

  refreshDives() {
    this.loading = true;
    this.errorMessage = null;
    this.dives = [];

    this.api.getDives()
      .then(data => {
        this.dives = data
        this.loading = false;
      }, errMsg => {
        this.errorMessage = errMsg;
        this.loading = false;
      });
  }
}
```

This construct uses the traditional promise operation syntax:

```
promiseOperation()
  .then(resolvedHandler, rejectedHandler)
```

We can use the method chaining syntax with the **catch()** function, too:

```
promiseOperation()
  .then(resolvedHandler)
  .catch(rejectedHandler)
```

I guess the second construct is easier to read:

```
this.api.getDives()
  .then(data => {
    this.dives = data
    this.loading = false;
  })
  .catch(errMsg => {
    this.errorMessage = errMsg;
    this.loading = false;
  });
```

Property Binding, Again

In connection to this sample, it is worth to mention property binding again. The template of **DiveLogComponent** conditionally shows or hides elements on the page: the progress ring and the error message. In *Chapter 2*, you already learned about the **ngIf** directive, which is a good candidate for such a task. Listing 4-7 shows that in this chapter, we use the **hidden** HTML attribute with property binding syntax.

Listing 4-7: dive-log.template.ts (Exercise-04-03)

```
<div class="container-fluid">
  <h1>My Latest Dives (Angular/TypeScript)</h1>
  <div class="row">
    <div class="col-sm-8" >
        <button class="btn btn-lg btn-primary"
          (click)="refreshDives()">
          Refresh
        </button>
        <img [hidden]="!loading" src="images/progressring.gif" />
      </div>
    <div class="col-sm-4">
      <input #searchBox class="form-control input-lg"
        placeholder="Search"
        (keyup)="0" />
    </div>
  </div>

  <div class="row" [hidden]="!errorMessage">
    <div class="col-sm-12">
      <h3 class="text-danger">{{errorMessage}}</h3>
    </div>
  </div>

  <div class="row">
    <div class="col-sm-4"
      *ngFor="let dive of dives | contentFilter:searchBox.value">
      <h3>{{dive.site}}</h3>
      <h4>{{dive.location}}</h4>
      <h2>{{dive.depth}} feet | {{dive.time}} min</h2>
    </div>
  </div>
</div>
```

As a result of these bindings, the progress ring—it is represented by a GIF image—is displayed only when the info is being loaded; the error message is shown if there is any error.

Behind the scenes, Angular evaluates the template expression of both **[hidden]** attributes and assigns the result to the **hidden** property of the DOM object that represents the corresponding element. Once again, **[hidden]** is not an Angular directive.

Using the Http Service

Using backend services is a common task in today's web applications. In the JavaScript world, this chore is undertaken by the **XMLHttpRequest** object, but most client libraries provide their toolset to encapsulate **XMLHttpRequest** in a way that fits the best for the particular framework.

Angular provides the **Http** service that implements several methods, such as **get()**, **post()**, **put()**, **delete()**, and a few others to carry out appropriate requests. The **Http** service supports a new style, the *asynchronous observable pattern*, by default, but we can use promises, too.

> **NOTE**: *The asynchronous observable pattern originates from the RxJS (Reactive Extensions, https://github.com/ReactiveX/RxJS) library, which is endorsed by Angular. In Chapter 11, you can learn more details about how Angular leverages this design pattern.*

Because we built the sample in **Exercise-04-03** on promises, we can easily turn it into an app that talks to a real backend. The **Http** service is implemented in **HttpModule**, and thus we import that module in **app.module.ts** (Listing 4-8). This import ensures that we can use Http service in any class within the application root module.

Listing 4-8: app.module.ts (Exercise-04-04)

```
import {NgModule} from '@angular/core';
import {BrowserModule} from '@angular/platform-browser';
import {HttpModule} from '@angular/http';

import {DiveLogComponent} from './dive-log.component';
import {ContentFilterPipe} from './content-filter.pipe';
import {DiveLogApi} from './dive-log-api.service';

@NgModule({
  imports: [
    BrowserModule,
    HttpModule
  ],
  declarations: [
    DiveLogComponent,
    ContentFilterPipe
  ],
  providers: [DiveLogApi],
  bootstrap: [DiveLogComponent]
})
export class AppModule { }
```

The new version of **DiveLogApi** leverages the **Http** service (Listing 4-9).

Listing 4-9: dive-log-api.service.ts (Exercise-04-04)

```
import {Injectable} from '@angular/core';
import {DiveLogEntry} from './dive-log-entry';
import {Http} from '@angular/http';
import 'rxjs/add/operator/toPromise';

@Injectable()
export class DiveLogApi {
  private DIVE_LOG_API_URL =
    'http://unraveling-ng.azurewebsites.net/api/backendtest/dives';

  constructor(private http: Http) {
  }

  getDives() {
    return this.http.get(this.DIVE_LOG_API_URL).toPromise()
      .then(resp => resp.json())
      .catch(err => {
        let errMsg = (err.message)
          ? err.message
          : err.status ? `${err.status}: ${err.statusText}` : 'Server
error';
        console.error(errMsg); // log to console instead
        return Promise.reject(errMsg);
      })
  };
}
```

Using the **Http** service that is injected into the constructor, **getDives()** invokes the **get()** method with the URL of the service endpoint. Because **Http** uses the asynchronous observable pattern, we need to use the **toPromise()** method to convert the **Observable** object retrieved by **get()** to a promise. The full RxJS library is large; Angular exposes only a minimal version. RxJS has many operators that can be used on **Observable**, one of them is **toPromise()**. To access it, we need to import the **rxjs/add/operator/toPromise** module.

If the call is successful, the promise is resolved with a **Response** object that is passed in the **resp** argument. The **json()** function converts the JSON body of the response into a JavaScript object—and with the URL specified in the code, it results in a **DiveLogEntry** array.

Should be there any error, the **catch()** method receives an error object and tries to obtain information from it. If this error does not have either a **message** or a **status** property, it returns with the "Server error" string.

> *HINT: Try to generate an error, for example, make a typo in the URL. Check how the app displays the error message.*

When you run the app, you can check that it leverages the remote backend service to get the list of dive log records. Figure 4-5 shows the JSON response coming back from the backend.

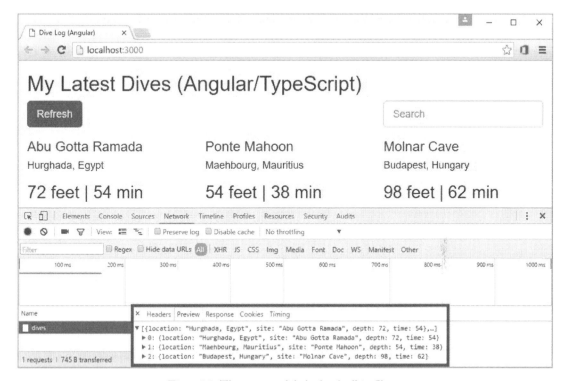

Figure 4-5: The response of the backend call in Chrome

A Few Remarks on the Backend Service

The backend service used in **Exercise-04-04** is a real service that runs on Microsoft Azure infrastructure. The service URL always retrieves a hard-coded collection of log entries. You can type (or copy) the service URL into your browser to check what it returns. Depending on the browser you use, it may retrieve either a JSON or an XML structure that represents the logbook entries.

For example, when you use it from Microsoft Edge or Internet Explorer, it retrieves a JSON response. Should you emit the call from Chrome, it would return an XML response that is shown in the browser (Figure 4-6).

When the **Http** service uses this URL with the get method, it passes this Accept header:

```
Accept: */*
```

The server at the backend can retrieve the response with JSON and XML encoding, where JSON takes priority over XML. Receiving the Accept header above, the server answers with JSON.

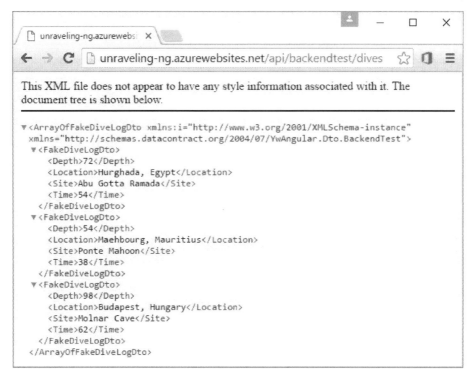

Figure 4-6: The XML response displayed in Chrome

There is another important thing the backend handles. The web application runs on your local machine (**http://localhost:3000**), but the backend service operates in another domain (**http://unraveling-ng.azurewebsites.net**). Usually, the browser would not accept the content because it comes from another domain (it is a cross-origin resource sharing scenario). The server retrieves a special header in the response, which lets the browser know that the content can be accepted, even if the request was served in a different domain:

```
Access-Control-Allow-Origin: *
```

> **NOTE**: *You can get more information about cross-origin resource sharing (CORS) here:* https://en.wikipedia.org/wiki/Cross-origin_resource_sharing.

REST Operations with the Http Service

As you saw, it is pretty easy to use the **Http** service. In real scenarios, you not only query data from the backend, but often add new information, modify existing data, or remove entities. In this section, you will learn a few more things about the **Http** service—through a compound example.

In Chapter 2 and 3, we built a dive site maintenance application with multiple views. In this section, we build another single page app that maintains dive locations information, and stores the data in a database on the backend.

The Starter Sample

In the **Exercise-04-05** folder, you find a fully functional starter version of the location management app that stores dive locations in memory. In a few steps, you will modify this app to use a remote backend data store through HTTP instead of a local one. Before going on, let's get acquainted with a few important details of the app.

When you start the app, it displays the list of dive locations, as shown in Figure 4-7. Here, you can edit or delete existing items, and even add new ones.

Figure 4-7: The list of dive locations

Unlike in the example, you created in Chapter 3, this UI leverages a single view—and thus, a single component—to provide all operations. Figure 4-8, Figure 4-9, and Figure 4-10 depict examples of these scenarios. You can cancel each operation, or commit them with the Add, Save, or Confirm buttons, depending on the particular function.

All operations mimic that it takes time while the backend answers the calls. While the client waits for the backend, a progress ring is displayed (Figure 4-11).

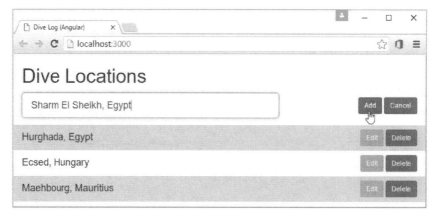

Figure 4-8: Adding a new location to the list

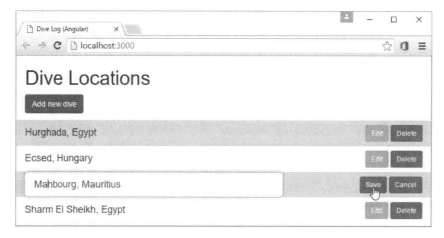

Figure 4-9: Modifying an existing location

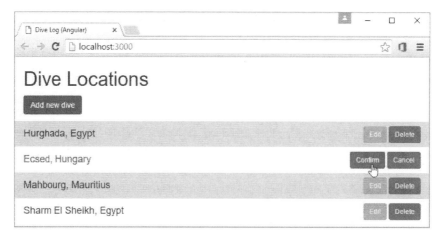

Figure 4-10: Deleting a location item

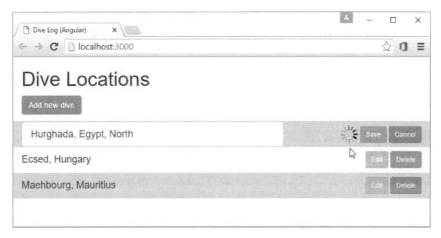

Figure 4-11: The progress ring indicates that an item is being saved

Implementing a Fake Backend Service

The most interesting part of the app is the implementation and consumption of the API that talks to the backend. The app's UI logic is implemented in a single component, **LocationsComponent**, which invokes backend operations through a method, **useBackend()** as shown in Listing 4-10.

Listing 4-10: locations.component.ts (Exercise-04-05)

```
import {Component} from '@angular/core';
import {Location} from './location';
import {LocationsApiService} from './locations-api.service'

@Component({
  selector: 'ywLocations',
  templateUrl: 'app/locations.template.html',
  // --- 'styles' omitted
})
export class LocationsComponent {
  // ...
  constructor(private api: LocationsApiService) {
    // ...
  }

  // ...
  add() {
    this.useBackend(-1, () =>
      this.api.addLocation({
        id: 0,
        displayName: this.locationName
      }))
  }
```

```
save() {
  this.useBackend(this.selectedId, () =>
    this.api.updateLocation({
        id: this.selectedId,
        displayName: this.locationName
    }))
}

remove(id) {
  this.useBackend(id, () =>
    this.api.removeLocation(id))
}
// ...
private useBackend(id: number, operation: () => Promise<any>) {
  this.busy(id);
  operation()
    .then(data => {
      this.refresh();
      this.complete(id);
    })
    .catch(err => {
      this.errorMessage = err;
      this.complete(id);
    });
}}
```

The **useBackend()** method accepts the id of the location to manage and an **operation** to invoke. The operation retrieves a promise, refreshes the page—or sets the error message to display—, and then removes the progress ring with the **complete()** method.

This example implements a fake **LocationsApi** that works in memory (Listing 4-11).

Listing 4-11: locations-api.service.ts (Exercise-04-05)

```
import {Injectable} from '@angular/core';
import {Location} from './location';

@Injectable()
export class LocationsApiService {
  private locations: Location[] = Location.StockLocations;

  getLocations() {
    return this.defer(100, () => this.locations.slice(0))
  }

  getLocationById(id: number) {
    return this.defer(10, () => {
      let itemId = this.getLocationIndexById(id);
        if (itemId >= 0) {
          return this.locations[itemId];
        } else {
```

```
          return null;
        }
     })
  }

  addLocation(location) {
    return this.defer(1000, () => {
      let newId = this.locations.length + 1;
      location.id = newId;
      this.locations.push(location);
    })
  }

  removeLocation(id) {
    return this.defer(1000, () => {
      let itemId = this.getLocationIndexById(id);
      if (itemId >= 0) {
        this.locations.splice(itemId, 1);
      }
    })
  }

  updateLocation(location) {
    return this.defer(1000, () => {
      let itemId = this.getLocationIndexById(location.id);
      if (itemId >= 0) {
        this.locations[itemId] = location;
      }
    })
  }

  private getLocationIndexById(id) {
    for (var i = 0; i < this.locations.length; i++) {
      if (this.locations[i].id == id) {
        return i;
      }
    }
    return -1;
  }

  private defer(time: number, operation: () => any) {
    return new Promise((resolve, reject) => {
      setTimeout(() => {
        let result = operation();
        resolve(result);
      }, time);
    })
  }
}
```

Each operation flows through the **defer()** method that delays them with the specified amount of time—with the help of a promise.

> *HINT: Take your time, play with the sample app and look into the details of the source code.*

Obtaining Your Access Token to the Backend

When I was preparing this book, I was thinking of a simple way to provide you a simple backend. First I conceived that I give you a Node.js implementation that you can install on your machine, but I abandoned that idea. I decided to provide you a backend in the cloud. In **Exercise-04-04** you used the **http://unraveling-ng.azurewebsites.net** backend—and this is the URL that you will use in this exercise, too, but this time with another service endpoint.

The API you have already used in **Exercise-04-04** was a cheap query operation that retrieved data from the backend. However, in this exercise, you are going to store data using the very same website with a different API. To allow multiple readers of this book to use the very same URLs hardcoded into the code samples—without interfering with each other's save operations—, I introduced a simple registration mechanism.

This mechanism requires only an email address from you, and no other information. The website uses this email address only to send you the User ID and the User Secret strings—the API requires these two pieces of information for authorization. You will not get any unsolicited emails to the specified address.

> *NOTE: You can use an invalid email address, but in this case, you won't receive an email with your User ID and User Secret information. You can still use the backend service by copying the required data from the screen.*
>
> *The system logs your activities. After 30 days of inactivity, your registration with all your example data will be removed without any prior notification.*

To get your user token, follow these steps:

1. Visit the **http://unraveling-ng.azurewebsites.net** page, and click the Register button, as shown in Figure 4-12.

2. You get to the registration page. Here, specify your email address. The dialog displays a code index (it's 56 in Figure 4-13). Lookup this code in Appendix A, which provides you the code value. Type (copy out) this code into the Code value field.

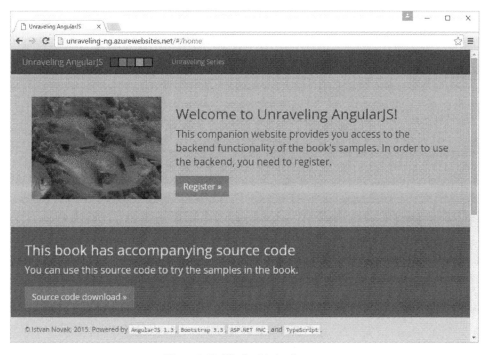

Figure 4-12: The book's landing page

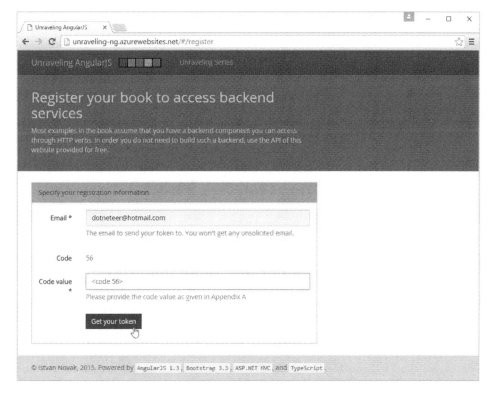

Figure 4-13: The registration form

3. Click the Get your token button. The system checks whether the code value you provided is valid and creates your User ID and User Secret, as a sample is shown in Figure 4-14. You can copy these values from the screen, and the information will be emailed to the address you provided, too.

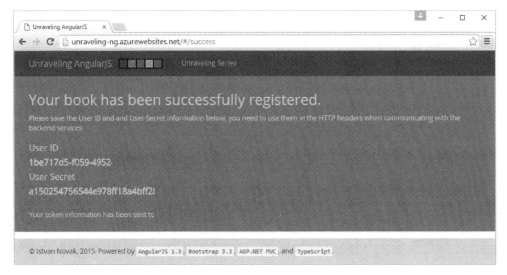

Figure 4-14: You are given a User ID and User Secret after successful registration

Invoking the Backend

With your access token, you have everything in your hand to use the **Http** service. Fortunately, the architecture used in the sample makes this refactoring easy: although you need to change the **LocationsApi** service completely, you do need to touch other components.

You can find the refactored solution in the **Execrcise-04-06** folder. Listing 4-12 shows the complete code of the new **LocationsApi** class.

Listing 4-12: locations-api.service.ts (Exercise-04-06)

```
import {Injectable} from '@angular/core';
import {Http, Headers, Request, RequestMethod, RequestOptions} from
'@angular/http';
import 'rxjs/add/operator/toPromise';

import {Location} from './location';

const URL = 'http://unraveling-ng.azurewebsites.net/api/dive/location/';
const USER_ID = 'b3f15a1b-8d0e-4e80-8b96-c36e66QQQQQQ';
const USER_SECRET =
'd6deac91eba6453bb6ad78c61e8052b616d2d830db20463ea6b4d8be918e29df6a0a4a1eb
dab46eb9cdd6b7ab3QQQQQQ'
```

```
@Injectable()
export class LocationsApiService {
  private locations: Location[] = [];

  constructor(private http: Http) {
  }

  getLocations() {
    let options = new RequestOptions({
      headers: this.getHeaders()
    });
    return this.http.get(URL, options).toPromise()
      .then(resp => resp.json())
      .catch(this.handleError)
  }

  getLocationById(id: number) {
    let options = new RequestOptions({
      headers: this.getHeaders()
    });
    return this.http.get(URL + id, options).toPromise()
      .then(resp => resp.json())
      .catch(this.handleError)
  }

  addLocation(location) {
    let options = new RequestOptions({
      headers: this.getHeaders()
    });
    return this.http.post(URL, location, options).toPromise()
      .then(resp => resp.json())
      .catch(this.handleError)
  }

  removeLocation(id) {
    let options = new RequestOptions({
      headers: this.getHeaders()
    });
    return this.http.delete(URL + id, options).toPromise()
      .then()
      .catch(this.handleError)
  }

  updateLocation(location) {
    let options = new RequestOptions({
      headers: this.getHeaders()
    });
    return this.http.put(URL, location, options).toPromise()
      .then()
      .catch(this.handleError)
  }
```

```
  private getHeaders() {
    return new Headers({
      'Authorization': `TenantSecret ${USER_ID},${USER_SECRET}`
    });
  }

  private handleError(err) {
    let errMsg = (err.message)
      ? err.message
      : err.status ? `${err.status}: ${err.statusText}` : 'Server error';
    console.error(errMsg);
    return Promise.reject(errMsg);
  }
}
```

Let's examine, how this class utilizes the **Http** service!

The code starts with importing all classes we use in the class. To add **Http** functionality, we need these imports:

```
import {Http, Headers, Request, RequestOptions} from '@angular/http';
import 'rxjs/add/operator/toPromise';
```

To emit an HTTP call, we create an HTTP request with authentication header, so we leverage the **Request**, **Headers**, and **RequestOptions** classes. As we did in **Exercise-04-04**, we use promises— and not **Observables**, as **Http** does it by default—, thus we import the **toPromise()** operator.

The backend refuses the call unless we provide a valid authorization token. We pass this token in the Authorization header using the User ID and User Secret values you obtained earlier:

```
...
const USER_ID = 'b3f15a1b-8d0e-4e80-8b96-c36e66QQQQQQ';
const USER_SECRET =
'd6deac91eba6453bb6ad78c61e8052b616d2d830db20463ea6b4d8be918e29df6a0a4a1eb
dab46eb9cdd6b7ab3QQQQQQ'

// ...
private getHeaders() {
  return new Headers({
    'Authorization': `TenantSecret ${USER_ID},${USER_SECRET}`
  });
}
...
```

Obviously, you need to change the **USER_ID** and **USER_SECRET** values to the ones you received during registration. Should you work with an invalid token, the backend would raise a 401 error (Figure 4-15).

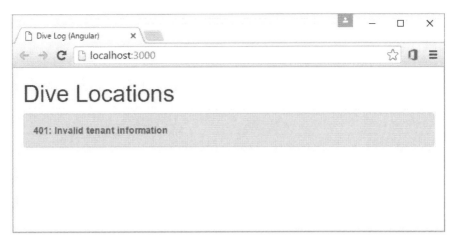

Figure 4-15: You cannot use the app with an invalid token

The **getHeader()** method creates the appropriate authorization header. If you intend to add other headers to each request, this method is the best place to do that.

Each service operation applies the same pattern as **getLocations()**:

```
getLocations() {
  let options = new RequestOptions({
    headers: this.getHeaders()
  });
  return this.http.get(URL, options).toPromise()
    .then(resp => resp.json())
    .catch(this.handleError)
}
```

The method's body starts with creating a new **RequestOptions** instance. The constructor accepts an object to set the URL, the HTTP method, headers, and other request parameters. In the code snippet above, we set only the headers. The real work is done by the **get()** method. We pass the service endpoint URL and the request options with the authorization header set. The **get()** operation immediately returns an **Observable** instance. We instantly convert it to a promise with the **toPromise()** operator.

As soon as the response comes back from the backend, we process it. If the call was successful, the promise is resolved, and the **then()** method is invoked, which extracts the list of locations from the response body and converts it into a **Locations[]** object with the **json()** function. If the status shows an error, the **handleError()** method extracts the information from the response body—using the same approach we applied in **Exercise-04-04**.

The **addLocation()**, **removeLocation()**, and **updateLocation()** methods follow this pattern. Obviously, they use the **post()**, **delete()**, and **put()** methods, respectively, which represent the HTTP verb with the same name. Although we do not utilize it, the **addLocation()** method retrieves the id of the newly added site:

```
return this.http.post(URL, location, options).toPromise()
  .then(resp => resp.json())
  .catch(this.handleError)
```

When you run the app, you can check the requests conveyed between the browser and the backend (Figure 4-16).

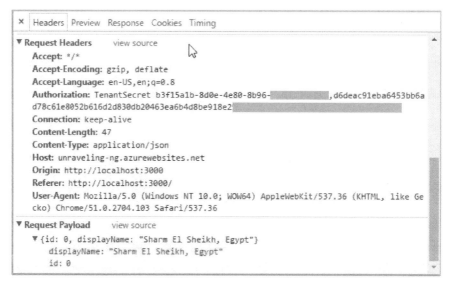

Figure 4-16: The POST request that adds a new location

Summary

Most web applications need to access backend services to retrieve and store information, carry out business processes. To use these services via the HTTP and HTTPS protocols, Angular offers you the **Http** service.

To access backend service endpoints, you must use asynchronous programming. By default, the **Http** service applies the asynchronous observable pattern, but you can leverage JavaScript promises with the help of the **toPromise()** operator and define resolved and rejected handler methods for all asynchronous calls.

In the next chapter, you will dive deeply into a few fundamental mechanisms of Angular.

Part II: Diving Deeper into Angular 2

Chapter 5: Bootstrapping, Templates, and Directives

WHAT YOU WILL LEARN IN THIS CHAPTER

Understanding the structure of index.html

Getting an overview of the bootstrap process

Using templates and bindings

Getting acquainted with the binding constructs

Understanding and creating structural and attribute directives

In Part I, you got acquainted with the fundamental Angular concepts, such as components, services, directives, styles, dependency injection and so on. Now, you know how you can leverage them in your apps.

In Part II, we dive deeper to learn not only what a particular constituent part of Angular does but also how it plays its role and what kind of mechanisms keep it working. In this chapter, we start at the very beginning and examine how an Angular app starts and how the framework uses components to implement an app with its UI and logic.

Understanding index.html

Though every sample we used so far displays the `index.html` page in the browser, we did not treat yet how this file contributes in an Angular app. Most exercises start with a simple `index.html` file like this:

```
<!DOCTYPE html>
<html>
<head>
  <title>Dive Site Maintenance</title>
  <link href="/node_modules/bootstrap/dist/css/bootstrap.min.css"
    rel="stylesheet" />
  <script src="/node_modules/core-js/client/shim.min.js"></script>
  <script src="/node_modules/zone.js/dist/zone.js"></script>
  <script src="/node_modules/reflect-metadata/Reflect.js"></script>
  <script src="/node_modules/systemjs/dist/system.src.js"></script>

  <script src="systemjs.config.js"></script>
```

```
  <script>
    System.import('app').catch(function(err){ console.error(err); });
  </script>
</head>
<body>
  <yw-app>Loading...</yw-app>
</body>
</html>
```

You already learned that the `<yw-app>` tag is the one Angular will use to attach your app through a component. But how does Angular know that it should do its job? Although `index.html` loads some scripts from the `node_modules` folder and does some magic with a file named `systemjs.config.js`, these file names give you no hint about when Angular is loaded, and how it bootstraps the application. It seems that Angular somehow leverages Node.js (that is where the `node_modules` folder comes from).

Most JavaScript libraries and frameworks require just one or a couple of `<script>` tags with file names that provide a clear hint about what they contain. Angular is a bit different.

Framework or Platform?

Although Angular calls itself a framework—and this is the term I have used and will use in this book—, Angular is continuously shifting into becoming a platform to develop apps for mobile and desktop, too.

Angular is not tied to the browser. You can utilize Angular even on the server side to pre-render the UI and provide a faster startup than you can achieve in the browser. Although we represent the UI with HTML markup, Angular makes it possible to render other markups, even native UI.

The flabbergasting things this short paragraph mentions deserve their dedicated books—I am sure they will be written and released as soon as the platform is matured enough. In this book, I do not dive into any of these great feats, I stay by the client side and HTML.

I brought this topic on because it gives you a clue why the `index.html` file and the Angular framework's loading mechanism is a bit more complicated than is looks it should be.

Modularity

Angular itself is modular. Moreover, it utilizes some third-party libraries. Certain modules are part of the core framework, and thus they are required. Other modules are optional; you load them only when your app needs them.

To create flexible, maintainable, easily refactorable apps, you should leverage on modules, too. TypeScript does an excellent job—as you have already seen in the exercises—, it allows you to create and consume modules while taking most of the work off from your shoulders.

Node.js and Packages

The Angular team uses the features of Node.js and Node Package Manager (**npm**) to install and maintain third-party packages as well as Angular itself. The **index.html** file starts with loading these scripts:

```
<script src="/node_modules/core-js/client/shim.min.js"></script>
<script src="/node_modules/zone.js/dist/zone.js"></script>
<script src="/node_modules/reflect-metadata/Reflect.js"></script>
<script src="/node_modules/systemjs/dist/system.src.js"></script>
```

Let's see what role they play!

As HTML5 evolves continuously, browsers support HTML5 features differently. Polyfills help to cope this situation because they can implement features—either established or proposed features of the HTML5 web standard.

The first package, **core-js**, adds essential features of ES2015 to the JavaScript global context (**window**). Because other packages do or may build their functionality on ES2015, we start with loading **core-js**. In the future when all browsers support ES2015 features, this package might be removed. According to my experiences, you may omit in when using Chrome, Firefox or Edge, but your app will fail with Internet Explorer 11 or earlier.

The second package is **zone.js**. This polyfill implements the Zone specification that defines an easy way of propagating context information across related asynchronous operations. It is required by Angular, as the framework uses asynchronous operations heavily. Without Zone, Angular would not be as efficient to paint only that patch of UI that was affected by a particular action.

The third package, **reflect-metadata**, is a polyfill, too. As you learned, Angular relies on metadata when running an app—this information helps to identify the role of TypeScript (ES2015) classes. To get the metadata for an individual class, Angular utilizes **reflect-metadata**.

The structure of metadata may change as the TypeScript compiler evolves; you need to use the version of **reflect-metadata** that works with the TypeScript package you use for development.

The fourth package, **Systemjs** is a universal module loader that complies with the ES2015 specification and allows loading modules dynamically. The world of module loaders is pretty diverse. There is about a half dozen of them that are frequently used by development teams.

Angular is tested heavily with **Systemjs**. You can use it with other module loaders, but if you are not very experienced and savvy with dynamic module loaders, it is the best to go on with this library.

How Systemjs Works

Systemjs can take care all the subtle things you need to carry out to find and load your apps' modules. To leverage these great features, you need to configure Systemjs once it has been loaded. This is the task of the **systemjs.config.js** file that is loaded in **index.html**:

```
...
<script src="systemjs.config.js"></script>
<script>
  System.import('app').catch(function(err){ console.error(err); });
</script>
...
```

The call of **System.import()** loads the "app" package. As a result of the configuration, it will load and run the transpiled form of **app/main.ts**. Should this activity fail, the related issue would be logged to the browser's console output.

Now, let's see the how Systemjs is configured. Listing 5-1 shows the content of **systemjs.config.js**. This file is the one that you can find in each exercise folder.

Listing 5-1: systemjs.config.js (Exercise-05-01)

```
...
System.config({
  paths: {
    // --- Here we define alias
    'npm:': 'node_modules/'
  },
  // --- Map Angular modules to their files
  map: {
    // --- All sample apps are within the 'apps' folder
    app: 'app',
    // --- Angular bundels we use in the book
    '@angular/core':
      'npm:@angular/core/bundles/core.umd.js',
    '@angular/common':
      'npm:@angular/common/bundles/common.umd.js',
    '@angular/compiler':
      'npm:@angular/compiler/bundles/compiler.umd.js',
    '@angular/platform-browser':
      'npm:@angular/platform-browser/bundles/platform-browser.umd.js',
    '@angular/platform-browser-dynamic':
      'npm:@angular/platform-browser-dynamic/bundles/platform-browser-
dynamic.umd.js',
    '@angular/http':
      'npm:@angular/http/bundles/http.umd.js',
    '@angular/router':
      'npm:@angular/router/bundles/router.umd.js',
    '@angular/forms':
      'npm:@angular/forms/bundles/forms.umd.js',
```

```
    // --- Other libraries we use in the book
    'rxjs': 'npm:rxjs'
  },
  // --- We define how packages should be loaded when no
  // --- filename and/or no extension is defined
  packages: {
    app: {
      main: './main.js',
      defaultExtension: 'js'
    },
    rxjs: {
      defaultExtension: 'js'
    }
  }
});
...
```

> **NOTE**: *The code in Listing 5-1 is wrapped into a JavaScrip immediately invoked function expression (IIFE). For the sake of simplicity, I omitted the function wrapper.*

The `System.config()` method call in the first line carries out the loader configuration. It accepts a configuration object. Systemjs supports several ways to set up `config`, depending on whether we use a transpiler (Babel, Traceur, TypeScript), bundle module files or other features. In Listing 5-1, we use the `paths`, `map` and `packages` configuration options—the majority of code lines prepares the values of these properties.

The `paths` property defines a simple alias; we map the "`node_modules/`" to a shorter name, `npm`. The `map` property stores maps for packages. Here, we provide maps for module aliases to a location or package. Locations are relative to `index.html` (this file loads the `systemjs.config.js` script).

The `packages` property provides mapping and local context information for packages. Here we specify that our apps' entry point is the `main.js` file. The `defaultExtension` describes the file extension to apply when it is not explicitly laid down for the corresponding module.

There is an extra trick in the samples of the book. The exercise folders are two folder levels deeper than `node_modules`, so we use the `bs-config.json` file to redirect the exercise level `node_modules` folders to the one in the samples root.

```
{
  "port": 3000,
  "server": {
    "routes": {
      "/node_modules": "../../node_modules"
    }
  }
}
```

Bootstrapping Angular Components

Loading **index.html** into the browser prepares everything to start an Angular app. After loading the polyfill files and Systemjs, the invocation of **System.import('app')** loads **app/main.js** to bootstraps the app. Angular is not loaded unless you bootstrap the application root module.

The **Exercise-05-01** folder contains a simple version of the dive log app, but its **main.ts** file comments out the bootstrap code (Listing 5-2):

Listing 5-2: main.ts (Exercise-05-01)

```
import {platformBrowserDynamic} from '@angular/platform-browser-dynamic';
import {AppModule} from './app.module';

//platformBrowserDynamic().bootstrapModule(AppModule);
```

When you run the app, the browser's network tool clearly shows that no Angular module is loaded—and the "Loading..." text indicates it, too (Figure 5-1).

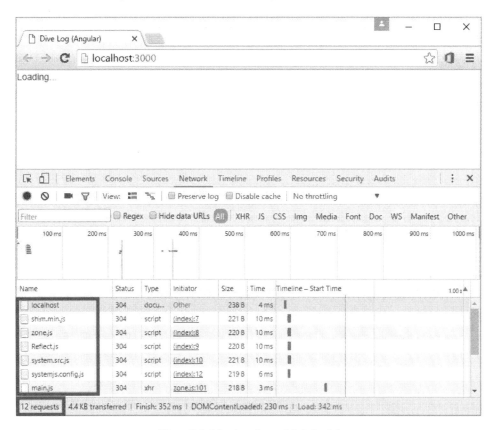

Figure 5-1: No Angular module is loaded

If you used plain JavaScript, importing the **@angular/platform-browser-dynamic** module would load Angular framework code into the browser. However, it does not happen in TypeScript because the compiler ignores any imported but unused modules. Compiling Listing 5-2 results in this JavaScript code:

```
"use strict";
//platformBrowserDynamic().bootstrapModule(AppModule);
//# sourceMappingURL=main.js.map
```

> **NOTE:** *Figure 5-1 shows that the app issued 12 requests, but the list displays only seven of them. The missing five requests are triggered by* **Lite-server** *to manage live browser synchronization.*

When you uncomment the **bootstrapModule()** call, the app loads about 30 Angular framework files, as Figure 5-2 indicates.

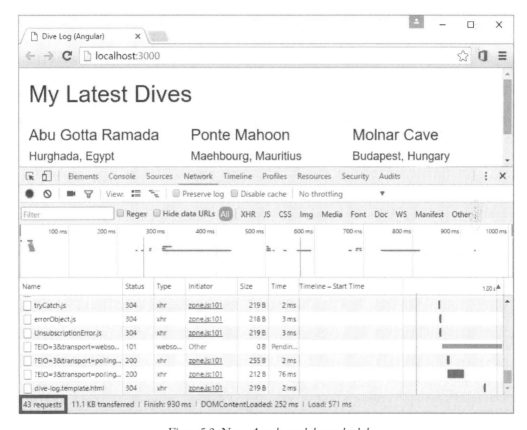

Figure 5-2: Now, Angular modules are loaded

> **NOTE:** *A few requests from the 43 in Figure 5-2 are initiated by the live browser synchronization and not Angular.*

The Bootstrap Process

When you call `bootstrapModule()`, Angular carries out these steps:

1. Using metadata of the module passed to `bootstrapModule()`—according to `@NgModule()`—, it reads the the `bootstrap` property to get the root application component. Though we can specify multiple components, for a moment assume that we have exactly one.

2. From the root component's metadata—according to `@Component()`—, the framework reads the value of the `selector` property to obtain the DOM element that represents the application root, and upgrades it into an Angular component.

3. The framework creates a new child injector for the component, the parent of which is the root injector. If you specify providers—you pass them in the second argument of `bootstrap()`—, the injector is initialized with those providers. You will learn more details about injectors and providers in *Chapter 6, Understanding Services and Dependency Injection*.

4. Each bootstrapped Angular component has its dedicated zone. A zone allows intercepting all asynchronous API in the browser and plays a crucial role in Angular change detections—as any change in Angular happens as a response to an external event coming in. In this step, the framework creates a zone for the component and connects it to the app's change detection domain instance.

5. The framework loads and parses the component template. The template is held as a separate DOM tree; it is loaded into the shadow DOM of the component. This shadow DOM is either native—provided the browser platform supports it—, or it is emulated by Angular.

6. The framework instantiates the component. It resolves dependencies—traverses through the entire dependency chain—and injects them into the component constructor.

7. When all the steps above are completed, Angular performs change detection to create and display the initial view for the component.

Bootstrap Issues

If there were any issue that would prevent the framework from completing these steps, it would not bootstrap the specified component. You observe such a situation from two signs: the view is not displayed—or the default "Loading..." text is shown—and there are error messages in the browser's console output.

Three typical issues frequently prevent components being bootstrapped properly:

#1. You define a particular component selector, but you do not add a corresponding HTML tag to the page (`index.html`) that host the component's view. For example, you mistype the component's selector:

```
// --- Component
@Component({
  selector: 'divelog',
  templateUrl: 'app/dive-log.template.html'
})
export class DiveLogComponent {
  //
}

// --- index.html
...
<myDivelog>Loading...</myDiveLog>
...
```

In such situations, Angular gives you an error message "*The selector **"selector-name"** did not match any elements*".

#2. Angular cannot resolve all dependencies to inject them into your component's constructor. In this case, you receive a "*No provider for **service-name!***" message.

#3. The component constructor raises an unhandled exception. An error message in the console output explicitly indicates it.

Step 5 of the bootstrap process detects the latter two issues, and though patches of the view might be displayed, the component does not work.

Component Selectors

Here are a few things worthy to know about the selectors of bootstrapped components:

You do not need to use an HTML element selector. Your component can be attached to a tag with an attribute selector:

```
// --- Component
@Component({
  selector: '[myDivelog]',
  templateUrl: 'app/dive-log.template.html'
})
export class DiveLogComponent {
  //
}

// --- index.html
...
<div myDivelog>Loading...</div>
...
```

You can put multiple HTML elements with the component selector into **index.html**, but Angular takes only the first of them into account and ignores the rest:

```
// --- Component@Component ({
  selector: 'divelog',
  templateUrl: 'app/dive-log.template.html'
})
export class DiveLogComponent {
  // ...
}
// --- index.html
...
  <divelog>Loading...</divelog>
  <divelog>Will be ingnored...</divelog>
  <divelog>This one too...</divelog>
...
```

Figure 5-3 demonstrates this plight.

Figure 5-3: Only the first matching element is bound to the component

> **NOTE**: *You can run the samples in the* **Exercise-05-02** *and* **Exercise-05-03** *folder, respectively, to check these features.*

Bootstrapping Multiple Components

You can bootstrap multiple components in a single web page, as shown in Listing 5-3.

Listing 5-3: app.component.ts (Exercise-05-04)

```
import {NgModule} from '@angular/core';
import {BrowserModule} from '@angular/platform-browser';

import {DiveLogComponent} from './dive-log.component';
import {PlannerComponent} from './planner.component';
```

```
@NgModule({
  imports: [BrowserModule],
  declarations: [
    DiveLogComponent,
    PlannerComponent
  ],
  bootstrap: [
    DiveLogComponent,
    PlannerComponent
  ]
})
export class AppModule { }
```

`PlannerComponent` is a simple component with its inline template and styles (Listing 5-4).

Listing 5-4: planner.component.ts (Exercise-05-04)

```
import {Component} from '@angular/core';

@Component({
  selector: 'planner',
  template: `
    <div class="container-fluid">
      <h1>My Dive Planner</h1>
      <h3>Under construction</h3>
    </div>
  `,
  styles: [`
    .container-fluid {
      background-color: #e0e0e0;
    }
  `]
})
export class PlannerComponent {
}
```

The `index.html` file contains the markup for both bootstrapped components:

```
...
<body>
  <divelog>Loading DiveLog...</divelog>
  <planner>Loading Planner...</planner>
</body>
...
```

Angular bootstraps both component and handles them separately. Because they run within the same browser window, they share singleton resources such as the title, location, cookies, and others.

If you go back to check the steps of the bootstrap process, it clearly indicates that these apps share only the root injector—or, with another name, the *platform injector*—but each has its zone, private injector, shadow DOM tree, and component instance.

Figure 5-4 demonstrates that they have even their dedicated CSS set: both components utilize the container-fluid CSS class, but the style rule highlighted in Listing 5-4 is applied only to its declaring component, `PlannerComponent`.

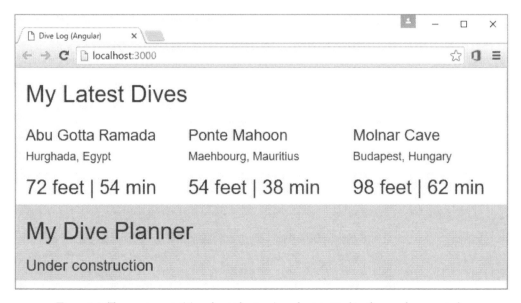

Figure 5-4: Two components (Angular applications) are bootstrapped in the same browser window

TypeScript

In this chapter, we deliberately forgot about that code examples are written in TypeScript, but the browser can run only JavaScript. To get our app ready to run, we need to compile the TypeScript code into JavaScript.

This is the task of the `tsc` (TypeScript Compiler) utility. When we run the `npm start` command to launch the app, it executes the following command line (as defined in `scripts` section of `package.json`):

```
concurrently "npm run tsc:w" "npm run lite"
```

The `concurrently` utility can run two or more commands simultaneously. In this case, it utilizes two `npm run` instructions to execute the `tsc:w` and `lite` scripts, which run `tsc -w`, and `lite-server`, respectively.

The `lite-server` utility runs a small development web server, opens the default browser to display the index (root) page. When it observes any file changes, it instructs the browser to refresh the current page.

TypeScript Compiler Options

The **tsc -w** command line starts the TypeScript compiler in *watch mode*: beside compiling all **.ts** file in the current folder and its subfolders, **tsc** observes the changes of TypeScript files and recompiles them.

To guide the compilation of **.ts** files, **tsconfig.json** defines settings and options. All exercises put this file into their root folder:

```
{
  "compilerOptions": {
    "target": "es5",
    "module": "commonjs",
    "moduleResolution": "node",
    "sourceMap": true,
    "emitDecoratorMetadata": true,
    "experimentalDecorators": true,
    "removeComments": false,
    "noImplicitAny": false
  }
}
```

As this JSON declaration suggests, the properties within **compilerOptions** define the settings **tsc** is expected to use. Table 5-1 describes the options used in our **tsconfig.json** file.

Table 5-1: tsconfig.json settings used in the book's exercises

Setting	Description
target	Specifies the ECMAScript target version. In this book, we assume that browsers are not entirely compatible with ES2015, so use the "**es5**" setting. For older browser versions, you can specify the "**es3**" option. If your browser supports ES2015 (formerly called ES6), set this option to "**es6**".
module	Declares the way the compiler generates module code. Here, with Angular, we use the Commonjs module pattern.
moduleResolution	Determines how to resolve modules. Most of the modules are managed by **npm**, so here we use the "**node**" module resolution style.
sourceMap	Generates **.map** files. You can use them to debug your app's TypeScript files in the browser.

Setting	Description
emitDecoratorMetadata	Emits design-type metadata for decorated declarations in the source
experimentalDecorators	Enables experimental support for ES7 decorators
removeComments	Removes all comments from the source, except copyright header comments beginning with /*!
noImplicitAny	By default (when set to false), the compiler silently defaults the type of a variable to **any** if it cannot infer the type based on how the variable is used. When this flag is true, and the TypeScript compiler cannot infer the type, it still generates the JavaScript files with reporting an error. Many developers and project teams prefer this setting because of stricter type checking.

> *NOTE: For a full list of available TypeScript compiler options and other settings, visit the TypeScript Handbook (http://www.typescriptlang.org/docs/handbook/tsconfig-json.html).*

Templates

In an Angular app, a component manages the logic of the app; a template represents the view the users see and interact with. As you already learned, templates are created in HTML, and they fully comply with the HTML syntax—even if it is not obvious at the first sight.

Angular templates use three important techniques that add expressiveness and power to the plain HTML markup: *interpolation, template expressions/statements*, and *bindings*.

Template Parsing

Angular has its built-in template parser that understands HTML and the template syntax. This parser is stricter than the browser. While the browser ignores several semantic errors to display the markup, the Angular template parser raises an error. For example, closing an **<h1>** tag with **</h2>** triggers the error message shown in Figure 5-5.

```
⊗ ▶ EXCEPTION: Template parse errors:
  Unexpected closing tag "h2" ("
      <h1>My Component[ERROR ->]</h2>
      "): MyComponent@1:20
```

Figure 5-5: Template parsing error

There are several HTML elements that are not useful in a template, like **<html>**, **<head>**, **<title>**, **<base>**, **<body>**, **<script>**, and others. Although the parser ignores them, you still need to use them syntactically properly to avoid error messages.

The parser builds an internal DOM according to the template. The ignored elements are omitted from this DOM. When displaying the UI, the DOM of the template is turned into a real DOM—that can be displayed in the browser—according to the logic of the corresponding component.

Interpolation

As you already learned, you can use the double-curly braces notation to file calculated strings into the text of HTML elements and attribute assignments. Take a look at this expression:

```
<p>Average dive time: {{ totalTime/diveCount }}</p>
<img src="profiles/{{ userName }}" heigth="50" width="50"/>
```

Interpolations wrap template expressions that are evaluated by Angular. Provided, the controller behind this template has set **totalTime**, **diveCount**, and **userName** to the values **6**, **4**, and **"joe"**, respectively, the following markup is displayed in the browser:

```
<p>Average dive time: 1.5</p>
<img src="profiles/joe" heigth="50" width="50"/>
```

You cannot use interpolation in HTML element or attribute names, so the highlighted constructs are invalid:

```
<{{myElement}}>My element</{{myElement}}>
<img {{myAttr}}="value" />
```

Template Expressions

When Angular meets with a template expression, it evaluates it in the context of the component the template is associated with. You can use Angular expressions in interpolations and property bindings. You can use JavaScript-like syntax when declaring template expressions, but you can use only a subset of operators and other JavaScript constructs.

When you create an expression, you should avoid side effects. For example, when reading a certain component property, you should avoid setting another property or value that is displayed by the template. For this reason, Angular does not allow applying the **new** operator, any assignment, decrement or increment operator in template expressions. Similarly, you cannot use the comma operator and separate more expressions with a semicolon.

JavaScript bitwise operators are not supported. One of them, the vertical bar ("|") has a new meaning; it is the *pipe operator*.

Beside restrictions, Angular defines a new operator, *safe navigation*. This operator is a great help in managing null values. Take a look at this code:

```
<h2>Diver: {{ diverName }}</h2>
<h3>Last dive: {{ dive.lastDive }} </h3>
```

If **diverName** is null, the expression within the first interpolation results in null, and a blank string is displayed to the right of "Diver:". However, if **dive** is undefined or null, the navigation to the **lastDive** property raises an error ("*TypeError: Cannot read property 'lastDive' of undefined*"), which is shown in the console log. The text of the **<h3>** tag is set to empty, so even the "Last dive:" literal remains undisplayed.

You can apply the safe navigation operator to avoid this issue:

```
<h2>Diver: {{ diverName }}</h2>
<h3>Last dive: {{ dive?.lastDive }} </h3>
```

When the reference at the left side of "**?.**" evaluates to **null** or **undefined**, the expression bails out without producing any error. This is a convenient way to avoid property path evaluation errors. You can even chain the safe evaluation operator with longer property paths:

```
<h3>Last dive year: {{ dive?.lastDive?.date?.year }} </h3>
```

Should any of **dive**, **lastDive**, or **date** be null or undefined, this expression would be evaluated without errors. This lengthy property path expression bails out when it hits the first null or undefined value.

It is not enough to emphasize only once that template expressions should be free from side effects. The code in Listing 5-5 creates a side effect by incrementing the **diveCount** property within the **avgTime()** function.

Listing 5-5: my-comp.component.ts (Exercise-05-05)

```
import {Component} from '@angular/core';

@Component({
  selector: 'my-comp',
  template: `
    <h2>Dive count: {{diveCount}}</h2>
    <h2>Average dive time #1: {{avgTime(diveCount)}}</h2>
    <h2>Average dive time #2: {{avgTime(diveCount)}}</h2>
  `
})
export class MyComponent {
  diveCount = 3;
  totalTime = 12;

  avgTime(count: number) {
    this.diveCount++;
```

```
    return this.totalTime/count;
  }
}
```

When you run this simple app, its output clearly demonstrates these side effects (Figure 5-6).

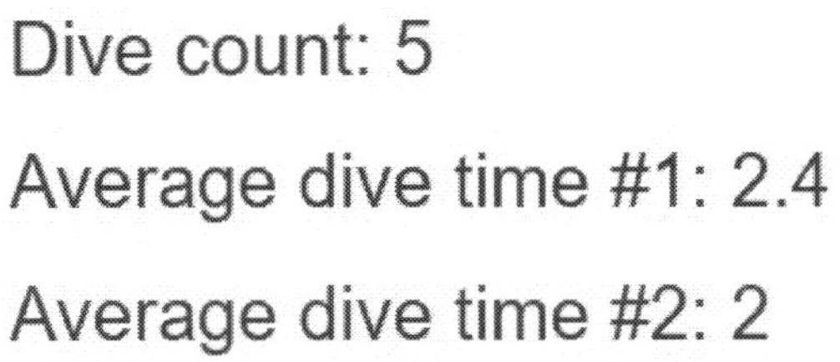

Dive count: 5

Average dive time #1: 2.4

Average dive time #2: 2

Figure 5-6: Side effects in template expressions

The value of **diveCount** is displayed to be 5, but the two calls of **avgTime(diveCount)** must get 5 and 6 as their input to produce 2.4 and 2. Strange, is not it? This phenomenon is the result of the side effect we put into **avgTime()**.

There are many events that may activate the change detection mechanism of Angular, and these can evaluate the template expressions more often than you think. Always strive to make these expressions quick; use caching whenever possible. Try to use short, simple template expressions. If you need to add complex expressions, create a property or function to get their value, and use the property or invoke the function in templates.

When creating the expressions, you can use only the members of the component that holds the template, and the names of *template reference variables*.

> **NOTE**: *You will learn about template reference variables soon.*

Template Statements

In templates, you can define actions that respond to events such as key presses, mouse movements, and so on. These responses can be defined as *template statements*.

```
<button class="btn btn-danger btn-lg"
  (click)="clearDives()">
  Clear dives
</button>
```

Just as template expressions, template statements use JavaScript-like syntax. Angular is less restrictive with statements, and it allows using the basic assignment ("="), the comma operator, and separating statements with semicolons. You still cannot use the **new**, increment, and decrement operators.

Just as expressions, statements are restricted to use only the members of the component that holds the template, and the names of template reference variables.

Understanding Bindings

Data binding is a powerful tool of Angular. With bindings, we can avoid writing imperative code that pushes values to and pulls values from HTML elements. Instead, we can use the declarative approach and delegate these tasks to the framework.

In Part I, you already met with data binding. Here is a short code snippet that demonstrates them:

```
<div class="row" *ngFor="let site of sites; let i=index"
  [style.background-color]="i%2 == 0 ? '#dddddd' : 'inherit'">
  <div class="col-sm-8">
    <h4 [style.color]="'maroon'">{{site.id}}: {{site.name}}</h4>
  </div>
  <div class="col-sm-4" style="margin-top: 5px;">
    <div class="pull-right">
      <button class="btn btn-warning btn-sm"
        (click)="edit(site)">
        Edit
      </button>
      <button class="btn btn-danger btn-sm"
        (click)="delete(site)">
        Delete
      </button>
    </div>
  </div>
</div>
```

In this code, all the highlighted parts represent data binding. If we would use imperative code to replace them, we would lose their expressiveness, and had to write more code. For example, the three data binding declared in the **<h4>** tag could be replaced with these actions:

#1: Get the **style** attribute value of the **<h4>** DOM element.

#2: Append the "**color: 'maroon'**" to this value, and store it back to the **style** attribute.

#3: Get the **site.id** value, concatenate it with "**: "** and the value of **site.name**.

#4: Store back the result to the **text** property of the **<h4>** DOM element.

Data binding is not a brand new technique; it is applied in many JavaScript frameworks such as Knockout, Ember, React, the previous version of Angular, and in others, too.

However, the way Angular manages bindings is a thing works mentioning. Syntactically it looks like as if Angular would set HTML attributes. Nonetheless, *Angular data binding works with the properties of DOM elements, components, or directives.*

This approach has a number of advantages over setting attribute values. The most important of them is that attribute values only initialize HTML elements, but they cannot be changed—unlike the properties of DOM. You can easily check this fact with the sample in Listing 5-6.

Listing 5-6: index.html (Exercise-05-06)

```
...
<body>
  <div class="container-fluid">
    <h1>Attribute vss Property</h1>
    <input id= "inp" type="text" value="John" onkeyup="onInputChanged()"/>
    <hr/>
    <p id="attr"></p>
    <p id="prop"></p>
  </div>
  <script>
    function onInputChanged() {
      var input = document.getElementById("inp");
      var attr = document.getElementById("attr");
      var prop = document.getElementById("prop");
      attr.textContent = "Attribute: " + input.getAttribute("value");
      prop.textContent = "Property: " + input.value;
    }
    onInputChanged();
  </script>
</body>
...
```

Here, the value attribute of **<input>** is initially set to "John". If the user of the app types "Jane" into the text box, the **value** *attribute* of **<input>** remains "John", though the **value** *property* of the DOM element is "Jane" (Figure 5-7).

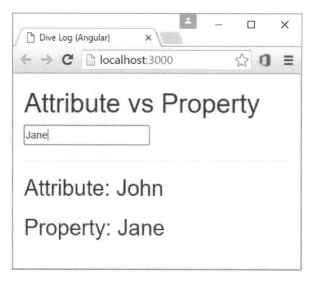

Figure 5-7: Attributes versus DOM properties

Angular support three categories of data binding types by the direction in which the data flows.

Property Binding

This type is a one-way data binding from the data source to the target. Value changes in a component or directive flow to the view or other components. You can define this kind of binding in three ways:

```
{{template expression}}
[target]="template expression"
bind-target="template expression"
```

As you see, string interpolation is a kind of property binding. Although "**[target]**" is a proper attribute name by the HTML5 standard, if your tools or conventions do not allow using it, you can apply the **bind-target** attribute name, instead. The target can be an element, component, or directive property.

With property binding, you can pass a value to the target, but you have no way to read the target property, nor to invoke a method on the target.

There is a fourth way of property binding, using the plain name of the property:

```
prop="literal value"
```

The framework assigns the string literal to the **prop** property of the DOM object of the corresponding target. Forgetting about the square brackets may lead to unexpected results, as Listing 5-7 demonstrates it.

Listing 5-7: my-comp.component.ts (Exercise-05-07)

```
import {Component} from '@angular/core';

@Component({
  selector: 'my-comp',
  template: `
    <h2 [class]="color">This is RED</h2>
    <h2 class="color">This is GREEN</h2>
  `,
  styles: [`
    .red {
      color: red;
    }

    .color {
      color: green;
    }
  `]
})
export class MyComponent {
  color = "red";
}
```

The first data binding expression uses `[class]`, so its value is a template expression that evaluates to "red". The second expression's value is taken into account as a string literal, "color". Thus, the first `<h2>` element applies the `.red` style rule, while the second utilizes `.color`. They have different colors, as shown in Figure 5-8.

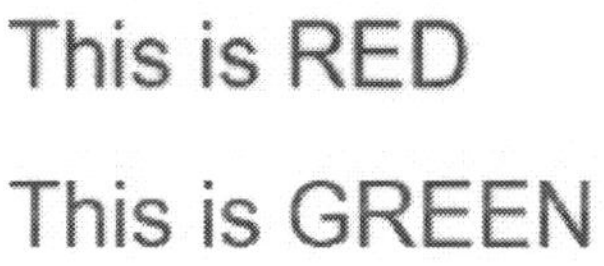

This is RED

This is GREEN

Figure 5-8: The two <h2> tags have different colors

Attribute, Class, and Style binding

There are special cases of property binding: *attribute*, *class*, and *style* binding.

There are several attributes—e.g. table span attributes, **SVG** or **ARIA** attributes—that do not correspond to DOM element properties, and so they cannot be set through DOM elements. *Attribute binding* provides a way to push values to these attributes, too:

```
[attr.targetAttribute]="template expression"
bind-attr.targetAttribute="template expression"
```

Class binding provides a convenient way to add or remove names to an element's **class** attribute. The syntax is similar to the attribute binding:

```
[class.className]="template expression"
bind-class.className="template expression"
```

The framework evaluates the template expression to a Boolean value. If it is true, it adds the specified *className* to the **class** attribute; otherwise, it removes it. Here is an easy-to-understand sample:

```
<div class="row" *ngFor="let site of sites; let f=first; let l=last"
  [class.topRow]="f"
  [class.bottomRow]="l">
  <!-- -->
</div>
```

Style binding makes it easy to set element style properties with a declarative way:

```
[style.styleProp]="template expression"
bind-style.styleProp="template expression"
```

The framework assigns the value given in the template expression to the **styleProp** style property. This code snippet sets alternating background color with style binding:

```
<div class="row" *ngFor="let site of sites; let i=index"
  [style.background-color]="i%2 == 0 ? '#dddddd' : 'inherit'">
  <!-- ... -->
</div>
```

Style binding adds some more syntax sugar. Those style properties that have units, can apply these units as extensions to the style:

```
[style.styleProp.unit]="template expression"
bind-style.styleProp.unit="template expression"
```

These samples help you understand how we can use these unit extensions:

```
<h2 [style.margin-top.px]="24">{{dive.site}}</h2>
<h3 [style.font-size.em]="1.5">{{dive.location}}</h3>
```

Event Binding

Event binding pushes data in one direction: from an element to a component or directive—so in the opposite direction as property binding does. When a user clicks a button or presses a key in an **<input>** element, these events can initiate actions at the component side with the help of event binding.

The event binding syntax is simple:

```
(eventName)="template statement"
```

```
on-eventName="template statement"
```

You can wrap the name of an event into parentheses, or use the canonical form with the "**on-**" prefix.

In **Exercise-01-08** we used this markup to create an event binding that tied the Add button to the **addDive()** component method:

```
<button class="btn btn-primary btn-lg"
  (click)="addDive()">
    Add new dive
</button>
```

We could have used the canonical form:

```
<button class="btn btn-primary btn-lg"
  on-click="addDive()">
    Add new dive
</button>
```

You should take care with event binding: if you use a wrong syntax or a mistyped event name, Angular will not throw an error—it does not respond to the event it does not understand. These definitions will *not* execute the **addDive()** method:

```
// Neither this...
<button class="btn btn-primary btn-lg"
  (clicked)="addDive()">
    Add new dive
</button>
// Nor this...
<button class="btn btn-primary btn-lg"
  (on-click)="addDive()">
    Add new dive
</button>
// ... and this does work, too
<button class="btn btn-primary btn-lg"
  on-clicked="addDive()">
    Add new dive
</button>
```

Events have parameters. When an event binding is declared, Angular creates its event handler and prepares an event object named **$event**. When the event is a native DOM element event, the **$event** is a standard DOM event object. Thus you can reach its properties, such as **type**, **target**, **bubbles**, and the others, and you can call its methods, such as **preventDefault()**.

The template statement can refer to **$event**. Listing 5-8 shows a simple component that uses the $event object to retrieve the name of the button clicked.

Listing 5-8: my-comp.component.ts (Exercise-05-08)

```
import {Component} from '@angular/core';

@Component({
  selector: 'my-comp',
  template: `
    <div class="container-fluid" style="margin-top:24px;">
      <button (click)="show($event)">Button #1</button>
      <button (click)="show($event)">Button #2</button>
      <button (click)="show($event)">Button #3</button>
      <h3>Clicked: {{clicked}}</h3>
    </div>
  `
})
export class MyComponent {
  clicked = "<none>";

  show(e) {
    this.clicked = e.target.textContent;
  }
}
```

The body of **show()** receives this event object through its **e** parameter, and can access the clicked button's text through **e.target.textContent** (Figure 5-9).

Figure 5-9: Using the $event object

Directives and components can define their events, and these custom events can be used with the event binding syntax just like standard DOM object events. We have already examined a few examples in *Chapter 2*, and you will find more advanced custom event samples in *Chapter 8*.

Two-Way Binding

Although Angular suggest using one-way bindings only, it has a special two-way binding mechanism to help to manage HTML forms. The Angular forms module defines a directive, **ngModel**, which allows you specify two-way binding with the "banana in the box" ("**[()]**") syntax:

```
<input [(ngModel)]="dive.location" />
```

Alternatively, if your tools or conventions do not allow using special characters in attribute names, you can use the canonical form:

```
<input bindon-ngModel="dive.location" />
```

When you develop data entry forms, you often need to both display a data property and update that property when the users make changes. The two-way binding is a syntax sugar for this purpose, and it works with any appropriate property, not just with **ngModel**. When using the two-way binding syntax, Angular transforms it behind the scenes as if you wrote this:

```
[propName]="expr"
(exprChange)="expr=$event"
```

Going back to the **<input>** sample, Angular turns the original syntax to this construct:

```
<input [ngModel]="dive.location"
       (ngModelChange)="dive.location=$event" />
```

This markup suggests that **dive.location** either must be a special object, or the **ngModelChange** event property must have an unusual event object. We do not reveal the secret of **ngModel** here, but you will learn all details in *Chapter 8*.

In my opinion, there are no real situations—except forms and data entry—where two-way bindings are practical. Do not use them—except with **ngModel**—unless you have a particular reason.

Template reference variables

We can place references to a DOM element or a directive within a template, and use these *template reference variables* in the DOM tree. To declare a reference variable, we declare attributes on a particular element, and prefix them with a hash ("**#**") or the **ref-** prefix:

```
<input #siteNameBox class="form-control input-lg" type="text"
  placeholder="site name"
  (keyup)="siteName=siteNameBox.value"
  (keyup.enter)="added()" />
```

When we utilize the template reference variable, we use its name without the declaration prefix, as **siteNameBox** demonstrates. The reference variable can be used within the entire template, even it can be referenced in a DOM element that precedes the declaration of the variable (Listing 5-9).

Listing 5-9: div-log.template.html (Exercise-05-09)

```
<div class="container-fluid">
  <h1>
```

```
    My Latest Dives
    <span *ngIf="searchBox.value">
       ({{searchBox.value}})
    </span>
  </h1>
  <div class="row">
    <div class="col-sm-4 col-sm-offset-8">
      <input #searchBox class="form-control input-lg"
        placeholder="Search"
        (keyup)="0" />
    </div>
  </div>
  <div class="row">
    <div class="col-sm-4"
      *ngFor="let dive of dives | contentFilter:searchBox.value">
      <h3>{{dive.site}}</h3>
      <h4>{{dive.location}}</h4>
      <h2>{{dive.depth}} feet | {{dive.time}} min</h2>
    </div>
  </div>
</div>
```

In this listing, we use the value of the `searchBox` template reference variable within the `<h1>` element. Nonetheless, it is declared in an `<input>` a few lines later. It still works as demonstrated in Figure 5-10—the search text is displayed in the heading.

When resolving names, the Angular templating engine looks up the host component definition of the template and the list of template reference variables, too.

Figure 5-10: Using a template reference variable

Directives

Components are great tools to create dynamic apps. When an app runs, Angular renders the components and transforms the corresponding DOM on-the-fly as the states of app components change. The recipes or instructions that determine how the DOM transformation is to carry out are called *directives*. According to their purpose and behavior, Angular has three kinds of directives:

Components. These are directives that have templates, and so manage their own patch of UI. When we build Angular applications, most of the time we design and create components.

Structural directives. These change the layout of the DOM either by adding or removing DOM elements. `ngIf`, `ngFor`, and `ngSwitch` are great examples. Structural directives do not have templates. We can create our custom structural directives, but we seldom need to do.

Attribute directives. Attribute directives can be attached to DOM elements, and although they do not have templates, they can change the appearance or behavior of elements they are annexed to.

In this section, we discuss structural and attribute directives.

Why We Need Structural Directives

In *Chapter 2*, you already learned that you could display or hide an HTML element or a component by setting the `hidden` DOM property to false, or true, respectively. The template definition in Listing 5-10 uses this technique to display diving rules for either daylight or night dives. Figure 5-11 shows the app in action.

Listing 5-10: dive-rules.template.html (Exercise-05-10)

```
<div class="container-fluid">
  <h1>Dive Instructions</h1>
  <button class="btn btn-primary"
    (click)="isNightDive=!isNightDive">
    Toggle dive type
  </button>
  <div class="row" [style.margin-top.px]="16">
    <div class="col-sm-12">
      <div [hidden]="!isNightDive">
        <h2>Daylight Dive Rules</h2>
        <ul>
          <li>Maximum depth: 60 feet</li>
          <li>Maximum time: 30 minutes</li>
          <li>Use torch</li>
          <li>Diving pairs need a backup torch, too</li>
        </ul>
      </div>
      <div [hidden]="isNightDive">
        <h2>Night Dive Rules</h2>
        <ul>
          <li>Maximum depth: 130 feet</li>
```

```
        <li>Maximum time: 60 minutes</li>
        <li>Turn back at half air!</li>
      </ul>
    </div>
  </div>
</div>
```

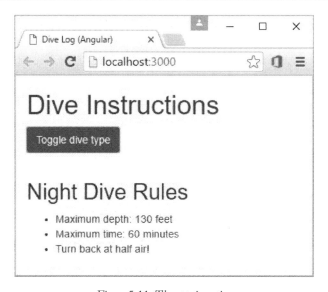

Figure 5-11: The app in action

With using **hidden**, both elements and their children remain the part of the DOM, though only one of them is visible (Figure 5-12).

```
▼ <div class="col-sm-12">
  ▼ <div hidden>  ⟸
      <h2>Daylight Dive Rules</h2>
    ▶ <ul>...</ul>
    </div>
  ▼ <div>  ⟸
      <h2>Night Dive Rules</h2>
    ▶ <ul>...</ul>
    </div>
  </div>
```

Figure 5-12: Both the displayed and hidden elements are in the DOM

Instead of the **hidden** attribute, you can use the **ngIf** directive (Listing 5-11).

Listing 5-11: dive-rules.template.html (Exercise-05-11)

```
...
<div class="col-sm-12">
```

```
    <div *ngIf="!isNightDive">
      <h2>Daylight Dive Rules</h2>
      <ul>
        <!-- ... -->
      </ul>
    </div>
    <div *ngIf="isNightDive">
      <h2>Night Dive Rules</h2>
      <ul>
        <!-- ... -->
      </ul>
    </div>
  </div>
  ...
```

Now, only that visible **`<div>`** element is in the DOM, as shown in Figure 5-13. As the **`ngIf`** condition value changes, the framework adds the visible element to the DOM and removes the invisible one.

The observable behavior of the app is the same with using either **`hidden`** or **`ngIf`**, but behind the scenes, different things happen.

When **`hidden`** is set to true, the corresponding element is not displayed, but the behavior of the component continues because the component remains attached to the hidden element. There may be events it listens to, and property value changes it responds. Moreover, Angular change detection is still watching for changes and updates the hidden element accordingly. Thus, even if being undisplayed, the hidden DOM element still consumes resources. Depending on its structure, the resource consumption may be subtle or, on the contrary, significant.

```
▼ <div class="col-sm-12">
    <!--template bindings={}-->
  ▼ <div>
      <h2>Daylight Dive Rules</h2>
    ▶ <ul>...</ul>
    </div>
    <!--template bindings={}-->
  </div>
  ::after
</div>
```

Figure 5-13: Only the visible element is in the DOM

When **`ngIf`** is applied to an element, the framework immediately removes the invisible element from the DOM, detaches any event listeners from it, stops checking changes and update its content. Elements are detached with all their children. Thus all resources held by children are freed, too.

To summarize, Angular structural directives provide a way to keep the layout of the DOM in the app's control by adding or removing DOM elements and thus allow influencing resource usage.

Creating Custom Structural Directives

Although the built-in Angular structural directives are perfectly adequate to implement complex layout, you can easily create your custom directives. The **Exercise-05-12** sample declares and applies two new structural directives, **ywDay**, and **ywNight**, respectively, which mark an element describing a daylight or night diving rule (Listing 5-12).

Listing 5-12: dive-rules.template.html (Exercise-05-12)

```html
<div class="container-fluid">
  <h1>Dive Instructions</h1>
  <h3>Dive type: {{diveType}}</h3>
  <button class="btn btn-primary"
    (click)="toggleDiveType()">
    Toggle dive type
  </button>
  <div class="row" [style.margin-top.px]="16">
    <div class="col-sm-12">
      <h2>Rules</h2>
      <ul>
        <li *ywNight="diveType">Maximum depth: 60 feet</li>
        <li *ywNight="diveType">Maximum time: 30 minutes</li>
        <li *ywNight="diveType">Use torch</li>
        <li *ywNight="diveType">Diving pairs need a backup torch, too</li>
        <li *ywDay="diveType">Maximum depth: 130 feet</li>
        <li *ywDay="diveType">Maximum time: 60 minutes</li>
        <li *ywDay="diveType">Turn back at half air!</li>
      </ul>
    </div>
  </div>
</div>
```

These directives work similarly to **ngIf**: they accept a string template expression and evaluate it. **ywDay** adds the corresponding element to the DOM if its value is "day"; otherwise, removes it. **ywNight** does the same but it checks for the "night" string.

Obviously, the two directives are very similar. Listing 5-13 shows **NightDirective** that implements **ywNight**.

Listing 5-13: yw-night.directive.ts (Exercise-05-12)

```typescript
import {Directive, Input} from '@angular/core'
import {TemplateRef, ViewContainerRef} from '@angular/core';

@Directive({
  selector: '[ywNight]'
})
export class NightDirective {
  constructor(
    private templateRef: TemplateRef<any>,
```

```
      private viewContainer: ViewContainerRef
    ) { }

  @Input() set ywNight(ruleType: string) {
    if (ruleType == 'night') {
      this.viewContainer.createEmbeddedView(this.templateRef);
    } else {
      this.viewContainer.clear();
    }
  }
}
```

Let's see how this code works!

Similarly to components, directives define the HTML element or attribute to which they are attached. In the code, the **@Directive()** annotation sets this selector to **[ywNight]**, and thus this directive is applied to any HTML elements that have a **ywNight** attribute.

The constructor of the class receives two objects via dependency injection. **TemplateRef** represents the template of the directive—you will learn about it soon—, **ViewContainerRef** is an object that can render the directive's template—that is to say, it can modify the DOM according to the template.

> *NOTE: You do not have to configure dependency injection to access **TemplateRef** and **ViewContainerRef**. After bootstrapping, Angular knows how to provide these types.*

The class uses the very same attribute, **ywNight**, with input property binding so that we can assign a template expression to the directive. We could use a separate attribute, but that would just bloat the directive. Because we need to add or remove the appropriate element to the DOM when the template expression's value changes, we declare a setter for **ywNight**.

Depending on the new value of **ywNight**, we add the template to the DOM by invoking the **createEmbeddedView()** method of the injected **ViewContainerRef**, or remove it with the **clear()** method. The code passes the template reference to **createEmbeddedView()**, so the rendering engine knows how to display it. The **clear()** method removes everything from the directive's view. Thus no visible DOM element will be rendered.

> *NOTE: The source code of **ywDay** is very similar to **ywNight**. You can find it in the **yw-day.directive.ts** file within the **Exercise-05-12** folder.*

To allow these directives work, we need to declare them in the application root module, as shown in Listing 5-14.

Listing 5-14: app.module.ts (Exercise-05-12)

```
import {NgModule} from '@angular/core';
```

```
import {BrowserModule} from '@angular/platform-browser';

import {DiveRulesComponent} from './dive-rules.component';
import {DayDirective} from './yw-day.directive';
import {NightDirective} from './yw-night.directive';

@NgModule({
  imports: [BrowserModule],
  declarations: [
    DiveRulesComponent,
    DayDirective,
    NightDirective
  ],
  bootstrap: [DiveRulesComponent]
})
export class AppModule { }
```

Because the **ywNight** and **ywDay** directives have the same structure, we can create a single class that manages both of them. Listing 5-15 shows the **RuleTypeDirective** class that carries out this job. The modified app can be found in the **Exercise-05-13** folder.

Listing 5-15: dive-rules.component.ts (Exercise-05-13)

```
import {Directive} from '@angular/core'
import { TemplateRef, ViewContainerRef } from '@angular/core';

@Directive({
  selector: '[ywDay],[ywNight]',
  inputs: ['ywDay', 'ywNight']
})
export class RuleTypeDirective {
  constructor(
    private templateRef: TemplateRef<any>,
    private viewContainer: ViewContainerRef
    ) { }

  set ywDay(ruleType: string) {
    this.createOrDestroy(ruleType == 'day');
  }

  set ywNight(ruleType: string) {
    this.createOrDestroy(ruleType != 'day');
  }

  createOrDestroy(create: boolean) {
    if (create) {
      this.viewContainer.createEmbeddedView(this.templateRef);
    } else {
      this.viewContainer.clear();
    }
  }
}
```

```
}
```

This code uses a very simple trick: it defines a compound selector to match the directive with both the **ywDay** and **ywNight** attributes. For the sake of demonstration, it introduces another form of declaring input properties—we can list their names in the **inputs** metadata property. To make the directive work as we expect, we have to set up a setter for both input properties.

Creating Compound Structural Directives

You probably observed that the directive syntax is a bit verbose, we need to assign the **diveType** template expression to each attribute:

```
...
<li *ywNight="diveType">Use torch</li>
<li *ywNight="diveType">Diving pairs need a backup torch, too</li>
<li *ywDay="diveType">Maximum depth: 130 feet</li>
<li *ywDay="diveType">Maximum time: 60 minutes</li>
...
```

The **Exercise-05-14** folder contains a version that fixes this issue. Instead of utilizing a single directive, it will apply two, as highlighted in Listing 5-16.

Listing 5-16: dive-rules.component.ts (Exercise-05-14)

```
...
<ul [ywRule]="diveType">
  <li *ywNight>Maximum depth: 60 feet</li>
  <li *ywNight>Maximum time: 30 minutes</li>
  <li *ywNight>Use torch</li>
  <li *ywNight>Diving pairs need a backup torch, too</li>
  <li *ywDay>Maximum depth: 130 feet</li>
  <li *ywDay>Maximum time: 60 minutes</li>
  <li *ywDay>Turn back at half air!</li>
</ul>
...
```

In this sample, the **RuleTypeDirective** class manages the **ywNight** and **ywDay** attributes, while **RuleDirective** is responsible for handling **ywRule**. The source code is more complex than the previous ones, for the two directive classes need to co-operate to render the appropriate view. In this scenario, **RuleTypeDirective** plays a passive role. Each attribute—**ywDays** and **ywNight**— registers its template and renderer information with **RuleDirective**. The lion's share of work is done by **RuleDirective** which renders the view by displaying the items that should be visible.

Listing 5-17 shows the source code of **RuleTypeDirective**.

Listing 5-17: yw-rule-type.directive.ts (Exercise-05-14)

```
import {Directive, Host} from '@angular/core'
```

```
import {TemplateRef, ViewContainerRef} from '@angular/core';
import {RuleDirective} from './yw-rule.directive';
import {RuleView} from './rule-view';

@Directive({
  selector: '[ywDay],[ywNight]',
  inputs: ['ywDay', 'ywNight']
})
export class RuleTypeDirective {
  constructor(
    private templateRef: TemplateRef<any>,
    private viewContainer: ViewContainerRef,
    @Host() private ruleDirective: RuleDirective
    ) { }

  set ywDay(ignored) {
    this.register(true);
  }
  set ywNight(ignored) {
    this.register(false);
  }

  register(isDay: boolean) {
    this.ruleDirective.registerView(isDay,
      new RuleView(this.viewContainer, this.templateRef));
  }
}
```

In this code, **RuleView** is a simple class to store the **TemplateRef** and **ViewContainer** instance we receive in the constructor. The **@Host()** annotation of the third constructor argument instructs Angular to pass the host element of the directives. Recall, it must be a **RuleDirective** instance that represents the **ywRule** element in the markup. Although **ywDay** and **ywNight** do not have template expressions, they will be assigned to **null** value because they are marked as input properties. We utilize this assignment in the setter methods to register the directives with **RuleDirective**, telling whether they represent a daylight or night dive rule.

RuleDirective is more complex, as you can see from its source code (Listing 5-18).

Listing 5-18: yw-rule.directive.ts (Exercise-05-14)

```
import {Directive, Host} from '@angular/core'
import {TemplateRef, ViewContainerRef} from '@angular/core';
import {RuleView} from './rule-view';

@Directive({
  selector: '[ywRule]',
  inputs: ['ywRule']
})
export class RuleDirective {
  private ruleValue: any;
```

```
private dayViews: RuleView[] = [];
private nightViews: RuleView[] = [];
private visibleViews: RuleView[] = [];

set ywRule(value: string) {
  if (value != 'night') {
    value = 'day';
  }
  if (this.ruleValue == value) return;
  this.ruleValue = value;

  this.refreshViews()
}

refreshViews() {
  this.removeVisibleViews();
  this.visibleViews = this.ruleValue == 'day'
    ? this.dayViews : this.nightViews;
  this.displayVisibleViews();
}

removeVisibleViews() {
  var visibleViews = this.visibleViews;
  for (var i = 0; i < visibleViews.length; i++) {
    visibleViews[i].destroy();
  }
}

displayVisibleViews() {
  var visibleViews = this.visibleViews;
  for (var i = 0; i < visibleViews.length; i++) {
    visibleViews[i].create();
  }
}

registerView(isDay: boolean, ruleView: RuleView) {
  if (isDay) {
    this.dayViews.push(ruleView);
  } else {
    this.nightViews.push(ruleView);
  }
  this.refreshViews();
}
}
```

The class stores the template and renderer information of its registered **ywDay** and **ywNight** children in several arrays—the array names suggest their usage—, **dayViews**, **nightViews**, and **visibleViews**, respectively.

When the template expression value **ywRule** is set, the setter method examines if there is a real change in the value, and if it is necessary, it refreshes the view.

The `refreshView()` method removes the views of the visible items—as they are going to be invisible—, and displays the views for either the daylight or night rules, depending on which of them becomes visible. On every occasion when the children of `ywRule` invoke `registerView()`, the view of visible children is refreshed.

Hidden and ngIf Revised

When comparing the operation of `hidden` and `ngIf`, we did not state and answer the question explicitly: which of them should be used? Well, it depends on.

Using `hidden` is faster, as the DOM is just slightly changed. However, it still uses the resources held by the invisible part of the DOM tree. Although `ngIf` is slower because it adds and removes subtrees in the DOM, it might be more frugal with Angular resources. Sometimes this behavior causes user experience issues because re-creating a new DOM element—when it is added to the DOM tree again—may consume tons of resources.

The `RuleDirective` sample points out to an important technique in regard to structural directives. Even if the DOM elements are removed, their templates and views should not be forgotten. `RuleDirective` stores this information, so it does not have to be re-created whenever the set of visible elements changes.

The built-in Angular directives are tuned for this kind of efficiency. All of them, including `ngIf`, `ngSwitch`, and `ngFor` take care of not changing the DOM unnecessarily. For example, when any variable part of the `ngIf` template expression changes, the DOM is modified only when the condition alters its previous value.

The template Element

There is an important question still floating over us. Why do we need the odd asterisk in front of `*ngIf`?

This notation is a syntax sugar. Behind the scenes, Angular transforms the element behind the directive into a `<template>` element. Assume, we have this markup snippet:

```
<site-list-view *ngIf="currentView == 'list'"
   [sites]="sites"
   (onAdd)="startAdd()">
</site-list-view>
```

Before further processing it, Angular expands it to this form:

```
<template [ngIf] ="currentView == 'list'">
   <site-list-view
      [sites]="sites"
      (onAdd)="startAdd()">
   </site-list-view>
</template>
```

The "*" prefix tells the framework to wrap the element *ngIf is applied to into <template> and create an ngIf property binding on the wrapper element with the original template expression.

Much ado about nothing, you can say. What do we gain with this structural change?

To be able to remove and later add the host element of *ngIf to the DOM, we must preserve its structure separately from the ngIf binding, and we need to keep the ngIf expression, too, so that we can detect changes in the condition value.

> **NOTE**: *We can use up to one directive with the asterisk syntax within a single HTML element. Just think it over, how would you wrap a single HTML element with multiple* <template> *tags?*

The Angular team has chosen the <template>-wrapped solution to handle structural directives. Not only ngIf, but any asterisk-prefixed directive is handled this way.

The HTML5 <template> tag is a great choice for this purpose. The standard defines it as "a mechanism for holding client-side content that is not to be rendered when a page is loaded but may subsequently be instantiated during runtime using JavaScript."

By default, a <template> is hidden in the browser, and its content is wrapped into a DocumentFragment DOM element.

Take a look at this markup snippet:

```
<template>
  <h1>This is not visible...</h1>
  <h2>...it is hidden, too</h2>
</template>
```

When the browser displays it, it transforms this markup to the DOM structure shown in Figure 5-14.

```
▼ <template>
  ▼ #document-fragment
      <h1>This is not visible...</h1>
      <h2>...it is hidden, too</h2>
  </template>
```

Figure 5-14: Rendering the <template> *tag*

Angular replaces the <template> tag with its DOM implementation.

In **Exercise-02-04**, the dive site maintenance application template used ngIf this way:

```
<site-list-view *ngIf="currentView == 'list'"
  [sites]="sites"
  (onAdd)="startAdd()">
</site-list-view>
<add-site-view *ngIf="currentView == 'add'"
```

```
  [siteId] = "newSiteId"
  (onAdded)="siteAdded($event)"
  (onCancel)="navigateTo('list')">
</add-site-view>
```

Angular transformed this markup to this DOM when the first **ngIf** condition was true and the List view is displayed:

```
<!--template bindings={
  "ng-reflect-ng-if": "true"
}-->
<site-list-view ...>
  <!--- Omitted for the sake of brevity -->
</site-list-view>
<!--template bindings={
  "ng-reflect-ng-if": "false"
}-->
```

Without going into details how the framework generated this fragment, let's observe the gist: Angular transformed the two **ngIf**-related **<template>** tags into DOM comment elements where the **bindings** object stores the value of the **ngIf** condition.

When the Add view is active, the generated DOM fragment changes to this:

```
<!--template bindings={
  "ng-reflect-ng-if": "false"
}-->
<!--template bindings={
  "ng-reflect-ng-if": "true"
}-->
<add-site-view ...>
  <!--- Omitted for the sake of brevity -->
</add-site-view>
```

This mechanism—among the others—provides that Angular can keep track of the current state of the app via the DOM.

> **NOTE**: *This short overview just showed the top of the iceberg. I will not go deeply into Angular implementation details, but many chapters of this book will provide you additional information.*

The ngFor Directive

Probably, **ngFor** is the most complex structural directive. The value of the directive is not a template expression or a template statement; it has its own *microsyntax* that the framework interprets. We have already met with this syntax:

```
<div class="row" *ngFor="let site of sites; let e=even" ...>
  <!-- ... -->
```

```
</div>
```

The core of this microsyntax is the "**let *localVar* of *iterable*"** expression. The framework traverses through each element within the *iterable* array and assigns that item to the **localVar** variable of the current template expression context. Thus the current iteration value is available in the corresponding template:

```
<div *ngFor="let dive of dives">
  <h3>{{dive.site}}</h3>
  <h4>{{dive.location}}</h4>
  <h2>{{dive.depth}} feet | {{dive.time}} min</h2>
</div>
```

> *NOTE: the iterable part of the microsyntax can be not only an array but a JavaScript iterable, too.*

ngFor exposes several values we can assign to local variables, as summarized in Table 5-2. These values support the iteration markup.

Table 5-2: ngFor exported values

Value	Description
index	Indicates the current loop iteration index—starting with zero.
first	This Boolean value indicates whether the current item is the *first* one in the iteration.
last	This Boolean value indicates whether the current item is the *last* one in the iteration.
even	This value is set to true to sign that the current index is an *even* number.
odd	This value is set to true to sign that the current index is an *odd* number.

We cannot use these values directly in the template of **ngFor**. We need to assign them to local variables, and use these locals to access the values in the template:

```
<div *ngFor="let site of sites; let e=even; let f=first; let l=last"
  [ngClass]="{ evenRow: e }"
  [class.topRow]="f"
  [class.bottomRow]="l">
  <!-- ... -->
</div>
```

As **ngFor** traverses through the iterable, it instantiates a template once for each item. These templates inherit the outer template context and the local variables set in the **ngFor** expression. For example, the markup snippet above can use the **site, e, f,** and **l** local variables beside the inherited template context.

Angular has no restrictions on the number of items to use in **ngFor**. However, you can imagine that long lists may have performance issues. Even if you have a large list, you generally change only a few items. If Angular recreated the templates of all items just because you inserted a new element or removed an old one, it would waste a lot of resources.

> *NOTE: I know, the term* list *is not precise enough, but for the sake of simplicity I use it instead of* array *or* iterable.

Angular has a more practical approach. It tracks each item—it assigns some identity information to items—and thus can recognize which items are affected by changes. Instead of rebuilding the entire set of templates—and modifying the DOM accordingly, **ngFor** propagates the changes with this approach:

When an item is added to the list, a new template instance is created for the new item, and this instance is added to the DOM. Similarly, when an item is removed, its corresponding template is removed from the DOM. When the order of the items changes, Angular uses their respective templates and reorders them in the DOM. Obviously, the framework keeps unchanged elements intact in the DOM.

Of course, in a single run of the change detection process, Angular may observe multiple changes within a list and combine these techniques to carry out all necessary DOM modification.

By default, Angular uses item references to check which items are modified within the list. Unfortunately, this cheap and simple technique does not work efficiently in every situation.

Assume, you query a long list of dive log entry items from the backend, and repeat this process several times within a minute. Probably the items obtained from the backend remain the same, or only a few of them changes. Because every new query creates totally new objects in the memory, Angular detects that each of them has been changed, and so it recreates all item templates and the corresponding DOM elements.

You can help Angular identify items with a *tracking function*, and thus avoid unnecessary recreation of templates.

The **Exercise-05-15** and **Exercise-05-16** folders contain two samples that demonstrate the benefits of a tracking function. Both samples emulate querying dive location information from a backend. While the first uses the default tracking of Angular, the second one utilizes a tracking function:

```
// --- Default implementation
// Template:
<div *ngFor="let item of gearItems">
```

```
  <!-- ... -->
</div>

// --- With tracking function
// Template:
<div *ngFor="let item of gearItems; trackBy:trackById">
  <!-- ... -->
</div>
// Component
...
trackById(index: number, item: GearItem) {
  return item.id;
}
...
```

You can assign a tracking function to **ngFor** with the **trackBy:*function*** syntax as shown in the code snippet. The tracking function—here, **trackById**—accepts two parameters, the **index** of the item, and the **item**, respectively. It is expected to return a unique value that identifies the item to track. This implementation returns the unique **id** of a **GearItem**.

The samples count the number of queries, the changes of the list, and the number of items changed. The Requery button gets the list from the backend, while the Reverse button reverses the list within the browser. Figure 5-15 and Figure 5-16 shows the two samples after clicking the Requery button three times and Reverse once.

Figure 5-15: Default tracking

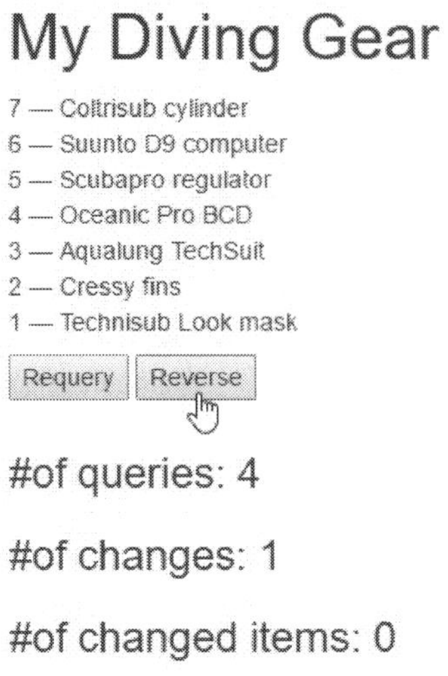

My Diving Gear

7 — Coltrisub cylinder
6 — Suunto D9 computer
5 — Scubapro regulator
4 — Oceanic Pro BCD
3 — Aqualung TechSuit
2 — Cressy fins
1 — Technisub Look mask

#of queries: 4

#of changes: 1

#of changed items: 0

Figure 5-16: Using a tracking function

Because the list was queried when the page was loaded, the "#of queries" counters displayed 4. With default tracking, each button click changed the list and all items within the list, thus, in total, 4 list changes and 28 item changes were detected. Figure 5-16 demonstrates that this app benefits from the tracking function. Because each item was tracked by its id, no item changes were detected. However, the Reverse button reversed the list. Thus a single list change was detected, for the items were reordered.

NOTE: Later, you will learn how these samples detected list and item changes.

Attribute Directives

Although we rarely need to create our custom structural directives, we can benefit a lot from implementing attribute directives. As you remember, attribute directives can be attached to DOM elements, and although they do not have templates, they can change the appearance or behavior of their host elements.

In *Chapter 3, Using Attribute Directives*, we already created an attribute directive. In this section, we create another one to learn more details. This directive, `SelectableDirective`, can be attached to any element to add hovering effect and allow selecting elements. Figure 5-17 shows that moving the mouse over an element allows changing its style—in this case, the background color.

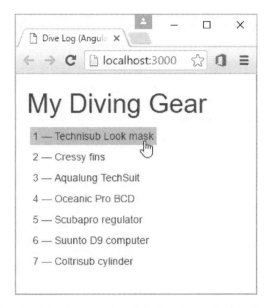

Figure 5-17: Implementing hover behavior with an attribute directive

The sample attribute directive allows clicking an element. This click toggles the "selected" state of the element. A selected element changes its background color and displays an exclamation mark in the front, as shown in Figure 5-18.

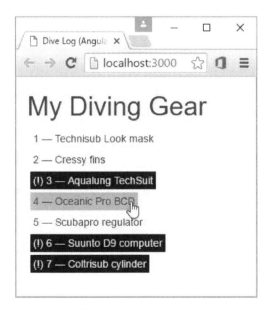

Figure 5-18: DOM modification with an attribute directive

The app is represented by the **DiveGearComponent**. Listing 5-19 shows the component's template. Each list item is represented by a **<gear-item>** tag that has applies **SelectableDirective** through the **ywSelectable** attribute.

Listing 5-19: dive-gear.template.html (Exercise-05-17)

```html
<div class="container-fluid">
  <h1>My Diving Gear</h1>
    <div *ngFor="let item of gearItems">
      <div class="row" style="padding: 4px;">
        <div class="col-sm-12">
          <gear-item [gear]="item"
            [ywSelectable]="'lightblue'" [ywId]="item.id"
            (onSelectionChanged)="selected($event)">
          </gear-item>
        </div>
      </div>
    </div>
</div>
```

Whenever the "selected" state of an item changes, the `selected()` method of
`GearItemComponent`—the host of `<gear-item>`—is invoked, which logs the event to the console
output.

Listing 5-20 shows the code of `SelectableDirective`.

Listing 5-20: selectable.directive.ts (Exercise-05-17)

```typescript
import {Directive, ElementRef} from '@angular/core';
import {Input, Output, EventEmitter} from '@angular/core';
import {HostListener, HostBinding} from '@angular/core';

@Directive({
  selector: '[ywSelectable]',
})
export class SelectableDirective {
  @Input('ywSelectable') hoverColor: string;
  @Input('ywId') itemId: number;
  @Output() onSelectionChanged = new EventEmitter<string>();
  isSelected: boolean;
  childSpan: HTMLElement;

  constructor(private element: ElementRef) {
    this.isSelected = false;
    this.childSpan = document.createElement('span');
    element.nativeElement.appendChild(this.childSpan);
  }

  @HostBinding('class.selected') get selected() { return this.isSelected;
}
  @HostBinding('style.padding.px') get extraPadding() { return 4; }

  @HostListener('mouseenter') onMouseEnter() {
    this.setAppearance(this.hoverColor, 'pointer');
  }
```

```
@HostListener('mouseleave') onMouseLeave() {
  this.setAppearance(null, null);
}

@HostListener('click') onClick() {
  this.isSelected = !this.isSelected;
  this.onSelectionChanged.emit(`${this.itemId}:${this.isSelected}`);
  this.childSpan.textContent = this.isSelected ? "(!) " : "";
}

setAppearance(color: string, cursor: string) {
  let style = this.element.nativeElement.style;
  style.backgroundColor = color;
  style.cursor = cursor;
}
}
```

Although the code looks a bit lengthy, it is pretty straightforward. Let's see its details!

The code utilizes two input properties, **hoverColor**, and **itemId**, respectively. We store **itemId** so that we can report back to the list the id of the item that has been selected or unselected. The **onSelectionChanged** event emits these notifications to the listening components:

```
@Input('ywSelectable') hoverColor: string;
@Input('ywId') itemId: number;
@Output() onSelectionChanged = new EventEmitter<string>();
```

The directive modifies the DOM of its host element. The constructor receives the host with the injected element attribute, and immediately adds a **<span>** element to represent the exclamation mark of the "selected" state:

```
constructor(private element: ElementRef) {
  this.isSelected = false;
  this.childSpan = document.createElement('span');
  element.nativeElement.appendChild(this.childSpan);
}
```

The **ElementRef** object that represents the host is a wrapper around the native DOM implementation. The constructor uses the standard JavaScript DOM library, and it utilizes the **nativeElement** property of the wrapped to access the DOM object.

The code applies the **@HostBinding()** decorator to attach directive properties to host properties:

```
@HostBinding('class.selected') get selected() { return this.isSelected; }
@HostBinding('style.padding.px') get extraPadding() { return 4; }
```

These bindings modify the host's class and style attributes through property binding.

With the **@HostListener()** decoration, we can bind event handler methods to the host element. The code declares handlers for the **mouseenter**, **mouseleave**, and click events respectively:

```
@HostListener('mouseenter') onMouseEnter() {
  this.setAppearance(this.hoverColor, 'pointer');
}

@HostListener('mouseleave') onMouseLeave() {
  this.setAppearance(null, null);
}

@HostListener('click') onClick() {
  this.isSelected = !this.isSelected;
  this.onSelectionChanged.emit(`${this.itemId}:${this.isSelected}`);
  this.childSpan.textContent = this.isSelected ? "(!) " : "";
}
```

Although this code is easy to read, we can improve it with metadata to make it more expressive. The **@Directive()** decorator allows us to get rid of the **@Input()**, **@Output()**, **@HostBinding()**, and **@HostListener()** annotations.

The **Exercise-05-18** folder contains the app with the modified **SelectableDirective**. It does not change the internals of the directive, but allows the reader of the code to get an immediate impression about what the class body contains by a single look at **@Directive()** (Listing 5-21).

Listing 5-21: selectable.directive.ts (Exercise-05-18)

```
...
@Directive({
  selector: '[ywSelectable]',
  inputs: [
    'hoverColor: ywSelectable',
    'itemId: ywId'
  ],
  outputs: ['onSelectionChanged'],
  host: {
    '[class.selected]': 'selected()',
    '[style.padding.px]': 'extraPadding()',
    '(mouseenter)': 'onMouseEnter()',
    '(mouseleave)': 'onMouseLeave()',
    '(click)': 'onClick()'
  }
})
export class SelectableDirective {
  // ...
}
```

The **inputs**, **outputs**, and **host** metadata properties provide an alternative to decorators. The **inputs** and **outputs** accept an array of strings that declare either a simple property name or—as used in this code—an aliased property name. The aliases used in **inputs** map the value of the

ywSelectable property to the **hoverColor** property of the class, the **ywId** to **itemId**. As the declaration of **outputs** show, the **onSelectionChanged** output property is mapped to the class member with the same name.

The **host** metadata property accepts an object where property names are keys to set up bindings; values are strings that represent templates expressions or statements.

A Few Comments on Directives

Directives are powerful Angular concepts that can be used to extend HTML with your custom elements and attributes. Nothing prevents you from using a directive to manage a complex patch of UI and even encapsulate business functionality.

However, the best way to use directives is to create utility recipes for your apps and components, and totally omit real business logic from them.

Although I treated them separately in this chapter, Angular directives should not be exclusively structural or attribute directives, you can mix these aspects within a single one.

Summary

In this chapter, you learned how to configure **Systemjs** to load Angular, and what does the framework do to bootstrap a component.

Components leverage templates and template binding syntax to express the connection between component classes and the UI they manage. Property, event, attribute, class and style bindings are all important concepts. Angular prefers unidirectional bindings but provides a syntax for two-way bindings, too.

The framework renders the components and transforms the corresponding DOM on-the-fly as the states of app components change. The recipes or instructions that determine how the DOM transformation is to carry out by directives. Components are directives, too.

In the next chapter, you will learn more details about **NgModules**, services, and the dependency injection mechanism.

Chapter 6: NgModules, Services, and Dependency Injection

WHAT YOU WILL LEARN IN THIS CHAPTER
Understanding the concept of Angular modules (NgModules)
Getting acquainted the ways we can use NgModules
Creating feature and shared modules
Understanding the concept of services and dependency injection
Getting an overview about the dependency resolution mechanism
Registering providers
Getting acquainted with several dependency injection scenarios

The bigger your apps are, the more files and entities you compose your application from. Sooner or later you need to find a way to manage this complexity before it overwhelms you. Here are a few typical issues you face when developing apps with huge number of co-operating parts:

#1. You need to struggle with finding where the source code of a certain component is.

#2. You need to manage dependencies among entities that belong to a particular business function.

#3. To keep the source manageable, you need to organize code files and components into clusters by several aspects.

#4. The structure of source code needs to be in synch with the app's features.

#5. You must manage the smart loading of modules: generally, you want to load only those modules into the memory that need to run certain features and intend to avoid loading the modules that are rarely used.

Angular offers several concepts to help you solve this issues. With `NgModules`, you can organize the code into logical modules, manage dependencies within an individual module, and among modules. You can extract classes, functions or values with a narrow, well-defined set of capabilities and form services from them. With Angular dependency injection, you can add these services to other application components, manage their configuration, and define the way service objects are instantiated.

In this chapter, you will learn the fundamental details of these concepts.

Understanding NgModules

In web development, there are many things called modules. Often, a JavaScript file that is loaded by a `<script>` tag is referred as a module, independently whether it uses any of the JavaScript module patterns (AMD, UMD, Node.js, Commonjs, or others). In TypeScript, we also have modules the concept of which overlaps with JavaScript.

The framework has a high-level concept—introduced as *Angular modules*—that helps organize an application and extend it with capabilities from external libraries. Angular leverages this concept in its internal structure and encourages you to build your apps with this approach.

To distinguish this definition of modules from the others, I will refer them as `NgModules`.

What an Angular Module (NgModule) Is

An Angular module is a container that consolidates pieces of an application or a library into a cohesive block of functionality. It is your design decision what you encapsulate into a module. It can be a collection of utilities, a set of application features, business functions, domain entities, and so on.

Applications can be composed of modules. Such an app can contain the modules developed especially for the app, others that are the part of the Angular frameworks, or even ones coming from third-parties. You may create modules for the mere purpose of being consumed by other third-party modules and apps.

A module is a class that is decorated with the `@NgModule()` annotation. Here is a sample:

```
import {NgModule} from '@angular/core';
import {BrowserModule} from '@angular/platform-browser';
import {FormsModule} from '@angular/forms';
import {HttpModule} from '@angular/http';

import {ChartsModule} from 'gigacompCharts';

import {AppComponent} from './app.component';
import {DashboardComponent} from './dashboard.component';
import {SiteManagementComponent} from './site-management.component';
import {DiveLogPipe} from './dive-log.pipe';

import {YwService} from './yw.service';
import {DiveLogApiService} from './dive-log-api.service'

import {CardDirective} from './card.directive'

@NgModule({
  imports: [
    BrowserModule,
    FormsModule,
    HttpModule,
```

```
      ChartsModule
    ],
    declarations: [
      AppComponent,
      DashboardComponent,
      SiteManagementComponent,
      DiveLogPipe,
      CardDirective
    ],
    providers: [
      YwService,
      DiveLogApiService
    ],
    bootstrap: [AppComponent]
})
export class AppModule { }
```

The **AppModule** class is here to provide an entity that can hold metadata. In this case, the metadata that is passed to the **@NgModule()** annotation. **AppModule** does not have any other role—its sole purpose is being the bearer of the module definition.

> *NOTE: As you already guessed out, I use the **NgModules** term after the **@NgModule()** annotation.*

The metadata describes the module. Each **@NgModule()** decoration declares several things:

#1. The components, directives, and pipes that belong together to provide the functionality of the module.

#2. The module-level services that the components, directives, and pipes *within* the component can use.

#3. The imports of other modules—Angular or third-party—to get access to the other components, directives, pipes, and services we intend to use within the module.

Each application has a root module—we call it *application root module*—which declares the components the application should bootstrap, too.

As you will learn later, modules can re-export objects they import.

The @NgModule Annotation

Let's take a look at the details of the sample above!

The **AppModule** class represents an application root module. We can infer it from the fact that its metadata has a **bootstrap** property signing that the app can be started by bootstrapping **AppComponent**:

```
@NgModule({
  // ...
  bootstrap: [AppComponent]
})
```

Because the module relies on other modules, it *imports* them:

```
@NgModule({
  imports: [
    BrowserModule,
    FormsModule,
    HttpModule,
    ChartsModule
  ],
  // ...
})
```

Importing a module means that the entities within **AppModule** can use the entities exported by the imported module. Thus, the app sees and may use any objects in **BrowserModule**, **FormsModule**, **HttpModule**, and **ChartsModule**. While the first three of them are the modules of the Angular framework, the fourth one is a (fictional) third-party module.

> **NOTE**: *A module imports other modules on which it depends. Those contain entities our module may utilize.*

At the beginning of the sample you see several import statements:

```
import {NgModule} from '@angular/core';
import {BrowserModule} from '@angular/platform-browser';
import {FormsModule} from '@angular/forms';
import {HttpModule} from '@angular/http';
// ...
```

These are not module imports! The code lines import the source files with the classes and variables that represent the external modules. You have to add these external definitions to the **NgModule** metadata to use them as imported modules.

AppModule has a section of declarations:

```
@NgModule({
  // ...
  declarations: [
    AppComponent,
    DashboardComponent,
    SiteManagementComponent,
```

```
    DiveLogPipe,
    CardDirective
  ]
  // ...
})
```

The array of objects passed to the **declarations** metadata property collects components, directives, and pipes that belong together and so must know about each other. To understand why it is important, let's assume that **AppComponent** has the following template:

```
<div>
  <h1>Younderwater Portal</h1>
  <yw-dashboard></yw-dashboard>
  <yw-site-management></yw-site-management>
</div>
```

Here, the **<yw-dashboard>** and **<yw-site-management>** elements are hosted by **DashboardComponent** and **SiteManagementComponent**, respectively. When parsing the template, Angular must observe that **<yw-dashboard>** and **<yw-site-management>** have host components; otherwise, it let the browser handle them as plain HTML tags.

The **declarations** metadata property provides support for **AppComponent** to understand what to do with these custom HTML elements. As the framework compiles the template, it checks the metadata of the entities passed to **declarations** and recognizes **DashboardComponent** and **SiteManagementComponent** as hosts. According to that information, **AppComponent** becomes aware of turning these elements into components.

> **NOTE**: *Of course, Angular traverses through the modules in* **imports** *to find hosts, too.*

The same situation may occur when we apply a pipe or a directive in a component template, so we need to add such pipe and directive classes to **declarations**.

The **@NgModule()** annotation contains a **providers** property, too:

```
@NgModule({
  // ...
  providers: [
    YwService,
    DiveLogApiService
  ],
  // ...
})
```

This metadata tells the framework that the application—of which **AppModule** the root module is—needs to provide **YwService** and **DiveLogApiService** when requested. According to this definition—as you will learn later in this chapter—, within **AppModule**, a singleton instance of

YwService and another singleton instance of **DiveLogApiService** will be injected to requesting entities.

BrowserModule

Angular application that run within the browser, must import **BrowserModule**. This core module encapsulates two other modules, **CommonModule**, and **ApplicationModule**.

CommonModule exports the common directives (such as **ngIf**, **ngFor**, **ngSwitch**, and others) and the core pipes (such as **DatePipe**, **NumberPipe**, and others—as you will learn in *Chapter 10*). **ApplicationModule** provides crucial services (such as the compiler, or the service responsible for initializing an app, among the others).

Modules and Module Hierarchies

So far, we have treated only application root modules. Real applications are rarely built from a single module. Rather, they are composed from other modules, as shown in Figure 6-1.

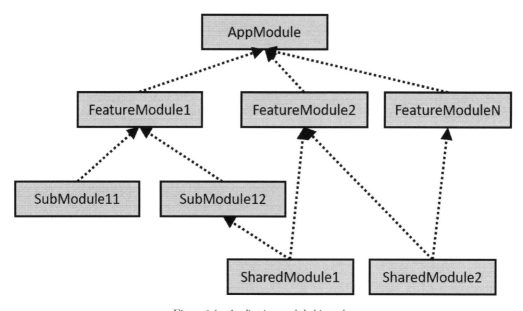

Figure 6-1: Application module hierarchy

In this figure, the arrows between modules represent imports: the head of each arrow points to the component that imports the other. The root **AppModule** utilizes three *feature modules*, **FeatureModule1**, **FeatureModule2**, and **FeatureModuleN**, respectively. These modules are imported either in the application root module or in other feature modules. As **SubModule11** and **SubModule12** indicates, feature modules can import other feature modules.

Observe, each feature module is imported exactly once!

There are modules—*shared modules*—that encapsulate functionality to be used in other modules (such as **SharedModule1** and **SharedModule2**). These are imported into one or more modules.

By default, every class encapsulated in a module—independently from its type—is taken into account as a hidden implementation detail. If you intend to publish any part of the module for external use, you have to *export* it.

Angular support *static* and *lazy loading* of modules. Every module we use in this chapter is statically loaded. In *Chapter 9, The Component Router*, you will learn how to manage a module with the lazy loading mechanism.

After so lengthy and dry theory, it is time to see code that plays with modules.

NgModules in Action

Let's create a skeleton app for the Younderwater Portal. The completed app will be assembled from five feature modules, and two shared modules; it displays the wireframe of the portal, as shown in Figure 6-2.

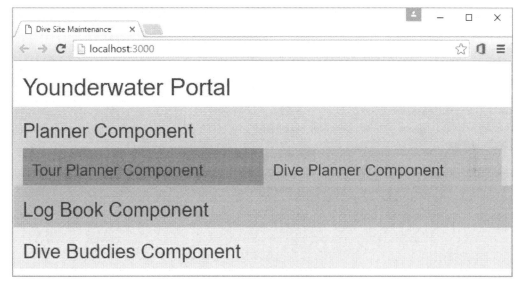

Figure 6-2: The wireframe of the portal

Each shaded rectangle in the figure represents a feature module; the two shared module contain utility directives. Figure 6-3 shows the module hierarchy and indicates the names of exported objects.

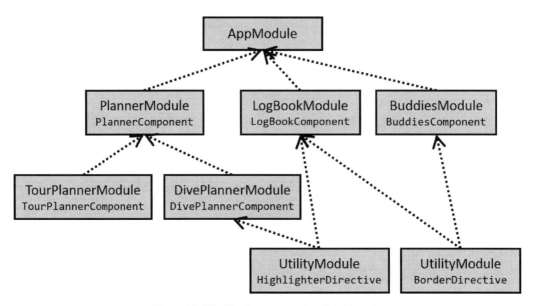

Figure 6-3: The Youderwater portal module hierarchy

Let's assume, we purchased the shared utilities from two third-parties, **GigaCorp** and **AcmeFactory**, respectively. Unfortunately, both of them put their directives in a module named **UtilityModule**, so somehow we need to resolve the naming conflict.

As our app will grow, it will contain dozens or hundreds of files. Before creating the skeleton, we need to establish a folder structure that does not constrain this expansion. Let's put all feature modules into their own folders that follow the hierarchy. For the sake of simplicity, we put both shared modules in their separate folders under **app**. Figure 6-4 shows the complete folder structure of the skeleton.

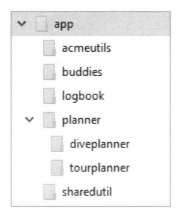

Figure 6-4: Folder structure

We build this app in a few phases. Let's start with creating the app and adding the first feature module, **PlannerModule**.

Adding PlannerModule

As the folder structure in Figure 6-4 shows, we put **PlannerModule**, and its single component, **PlannerComponent**, into the **app/planner** folder. The source code of **PlannerComponent** is straightforward (Listing 6-1).

Listing 6-1: planner.component.ts (Exercise-06-01)

```
import {Component} from '@angular/core';

@Component({
  selector: 'yw-planner',
  template: `
    <div class="row">
      <div class="col-sm-12">
        <h2>Planner Component</h2>
      </div>
    </div>
  `,
  styles: [
    `
    .row {
      background-color: #e0e0e0;
    }
    `]
})
export class PlannerComponent {
}
```

In **PlannerModule**, we need to add **PlannerComponent** to the **declarations**. Nonetheless, it is not enough to make the component available in the feature module. We need to export **PlannerComponent**; otherwise, Angular takes it into account as an internal implementation detail of the module. Listing 6-2 shows that we use the **exports** metadata property of **@NgModule()** to sign this intention.

Listing 6-2: planner.module.ts (Exercise-06-01)

```
import {NgModule} from '@angular/core';
import {BrowserModule} from '@angular/platform-browser';

import {PlannerComponent} from './planner.component';

@NgModule({
  imports: [BrowserModule],
  declarations: [PlannerComponent],
  exports: [PlannerComponent]
})
export class PlannerModule { }
```

We need to import **PlannerModule** into **AppModule** so that we can use **PlannerComponent** (Listing 6-3).

Listing 6-3: app.module.ts (Exercise-06-01)

```
import {NgModule} from '@angular/core';
import {BrowserModule} from '@angular/platform-browser';

import {AppComponent} from './app.component';
import {PlannerModule} from './planner/planner.module';

@NgModule({
  imports: [
    BrowserModule,
    PlannerModule
  ],
  declarations: [AppComponent],
  bootstrap: [AppComponent]
})
export class AppModule { }
```

Now, we can add the **<yw-planner>** tag to the **AppComponent** template (Listing 6-4). Because **AppModule** imports **PlannerModule**, the framework can find **PlannerComponent**, and from its metadata, it infers that the component is bound to **<yw-planner>**.

Listing 6-4: app.component.ts (Exercise-06-01)

```
import {Component} from '@angular/core';

@Component({
  selector: 'yw-app',
  template: `
    <div class="container-fluid">
      <h1>Younderwater Portal</h1>
      <yw-planner></yw-planner>
    </div>
  `
})
export class AppComponent {
}
```

Run the app and check that it works as expected. As Figure 6-5 shows, the portal displays the UI of the **PlannerComponent**.

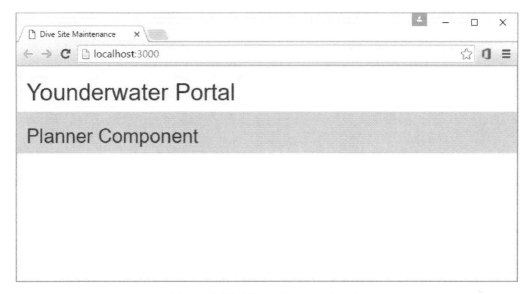

Figure 6-5: The PlannerComponent in action

Adding Feature Modules to PlannerModule

PlannerModule encapsulates two other feature modules, **TourPlannerModule** and **DivePlannerModule**. Each of them is in its folder under **app/planner** and they have very similar structure—for the sake of simplicity. While Listing 6-5 shows the source of **TourPlannerComponents**, in Listing 6-6, you can see that **TourPlannerModule** uses the same technique as its parent, **PlannerModules**, to encapsulate and export the main module component.

Listing 6-5: tour-planner.component.ts (Exercise-06-02)

```
import {Component} from '@angular/core';

@Component({
  selector: 'yw-tour-planner',
  template: `
    <div class="col-sm-6">
      <h3>Tour Planner Component</h3>
    </div>
  `,
  styles: [
    `
      .col-sm-6 {
        background-color: #00e0ff;
      }
    `]
})
export class TourPlannerComponent {
}
```

Listing 6-6: tour-planner.module.ts (Exercise-06-02)

```
import {NgModule} from '@angular/core';
import {BrowserModule} from '@angular/platform-browser';

import {TourPlannerComponent} from './tour-planner.component';

@NgModule({
  imports: [BrowserModule],
  declarations: [TourPlannerComponent],
  exports: [TourPlannerComponent]
})
export class TourPlannerModule { }
```

DivePlannerModule and its main component, **DivePlannerComponent** follows the same pattern. To make these subfeatures available in **PlannerModule**, we need to import both of the subfeature modules, as shown in Listing 6-7.

Listing 6-7: planner.module.ts (Exercise-06-02)

```
import {NgModule} from '@angular/core';
import {BrowserModule} from '@angular/platform-browser';

import {PlannerComponent} from './planner.component';
import {TourPlannerModule} from './tourplanner/tour-planner.module';
import {DivePlannerModule} from './diveplanner/dive-planner.module';

@NgModule({
  imports: [
    BrowserModule,
    TourPlannerModule,
    DivePlannerModule
  ],
  declarations: [PlannerComponent],
  exports: [PlannerComponent]
})
export class PlannerModule { }
```

Because these two subfeatures are closed into **PlannerModule**, and they are not supposed to be available outside of the host module, we do not need to export them. It would not lead to any error if we made them available with the **exports** metadata property. However, we do not want to expose them, as they are internal implementation details of **PlannerModule**.

Now, in the template of **PlannerComponent**, we can refer to these new components (Listing 6-8).

Listing 6-8: planner.component.ts (Exercise-06-02)

```
import {Component} from '@angular/core';

@Component({
  selector: 'yw-planner',
  template: `
    <div class="row">
      <div class="col-sm-12">
        <h2>Planner Component</h2>
        <yw-tour-planner></yw-tour-planner>
        <yw-dive-planner></yw-dive-planner>
      </div>
    </div>
  `,
  styles: [
    `
    .row {
      background-color: #e0e0e0;
    }
    `]
})
export class PlannerComponent {
}
```

When you run the app, proves that the two new components are integrated into the planner feature (Figure 6-6).

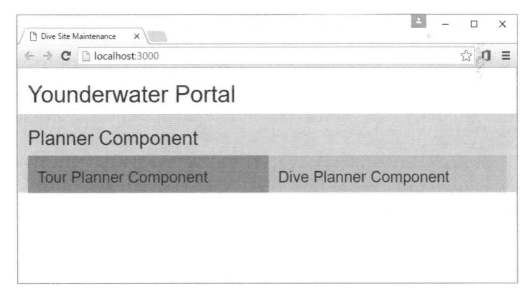

Figure 6-6: PlannerModule with its submodules

Adding LogBookModule and BuddiesModule

I suppose, without showing you the list of the two other feature modules, you can imagine that they follow the same pattern as **PlannerModule**. After adding them to the app, **AppModule** includes all the five feature modules: three of them—**PlannerModule**, **LogbookModule**, and **BuddiesModule**—directly, while **TourPlannerModule** and **DivePlannerModule** indirectly through **PlannerModule** (Listing 6-9).

Listing 6-9: app.module.ts (Exercise-06-03)

```
import {NgModule} from '@angular/core';
import {BrowserModule} from '@angular/platform-browser';

import {AppComponent} from './app.component';
import {PlannerModule} from './planner/planner.module';
import {LogBookModule} from './logbook/log-book.module';
import {BuddiesModule} from './buddies/buddies.module';

@NgModule({
  imports: [
    BrowserModule,
    PlannerModule,
    LogBookModule,
    BuddiesModule
  ],
  declarations: [AppComponent],
  bootstrap: [AppComponent]
})
export class AppModule { }
```

Listing 6-10: app.component.ts (Exercise-06-03)

```
import {Component} from '@angular/core';

@Component({
  selector: 'yw-app',
  template: `
    <div class="container-fluid">
      <h1>Younderwater Portal</h1>
      <yw-planner></yw-planner>
      <yw-log-book></yw-log-book>
      <yw-buddies></yw-buddies>
    </div>
  `
})
export class AppComponent {
}
```

The template of **AppComponent** includes the HTML tags for the two new feature component (Listing 6-10), and thus the portal's skeleton displays every block of the UI (Figure 6-7).

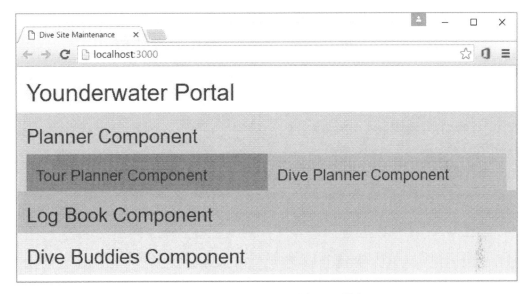

Figure 6-7: All feature modules are added

Extending the App with Shared Modules

Beside the feature modules, the Younderwater portal leverages two shared modules. Let's assume, we purchased them from software component vendors. For the sake of simplicity, these modules contain one-one directives. One vendor, GigaCorp, provides the **HighlighterDirective** (Listing 6-11), while the other, AcmeFactory delivers **BorderDirective** (Listing 6-12).

Listing 6-11: sharedutil/utility.module.ts (Exercise-06-04)

```
import {NgModule} from '@angular/core';
import {BrowserModule} from '@angular/platform-browser';

import {HighlighterDirective} from './highlighter.directive';

@NgModule({
  imports: [BrowserModule],
  declarations: [HighlighterDirective],
  exports: [HighlighterDirective]
})
export class UtilityModule { }
```

Listing 6-12: acmeutils/utility.module.ts (Exercise-06-04)

```
import {NgModule} from '@angular/core';
import {BrowserModule} from '@angular/platform-browser';

import {BorderDirective} from './border.directive';

@NgModule({
  imports: [BrowserModule],
  declarations: [BorderDirective],
  exports: [BorderDirective]
})
export class UtilityModule { }
```

When the mouse moves over a patch of UI, **HighlighterDirective** changes the background color, while **BorderDirective** draws a border. **DivePlannerComponent** uses **HighlighterDirective**, so the component's module, **DivePlannerModule** imports GigaCorp's **UtilityModule** from the **sharedutils** folder (Listing 6-13).

Listing 6-13: dive-planner.module.ts (Exercise-06-04)

```
import {NgModule} from '@angular/core';
import {BrowserModule} from '@angular/platform-browser';

import {DivePlannerComponent} from './dive-planner.component';
import {UtilityModule} from '../../sharedutil/utility.module';

@NgModule({
  imports: [
    BrowserModule,
    UtilityModule
  ],
  declarations: [DivePlannerComponent],
  exports: [DivePlannerComponent]
})
export class DivePlannerModule { }
```

To use one of the directives, it is enough to import the corresponding host module. Whenever we use the HTML attribute that matches with the directive's selector, Angular will find the host component by scanning the metadata of imported modules. Thus, we can use **HighlighterDirective** in **DivePlannerComponent**'s template without even referencing the directive (Listing 6-14).

Listing 6-14: dive-planner.component.ts (Exercise-06-04)

```
import {Component} from '@angular/core';

@Component({
  selector: 'yw-dive-planner',
```

```
  template: `
    <div class="col-sm-6"
      gc-highlight="red">
      <h3>Dive Planner Component</h3>
    </div>
    `,
  styles: [
    `
      .col-sm-6 {
        background-color: #ffe000;
      }
    `]
})
export class DivePlannerComponent {
}
```

> **NOTE**: *I added the same* `gc-highlight="red"` *attribute to the template of* `TourPlannerComponent`. *Because that component does not know* `HighlighterDirective`—*it does not import the directive's host module*—*the* `gc-highlight` *attribute is not associated with the directive. When you run the app, moving the mouse over the tour planner UI part will not change the background color, unlike the dive planner part, which does.*

`LogBookComponent` applies both third-party directives, so its module, `LogBookModule` must import the host modules of both directives. There is a little issue with them: both modules are named `UtilityModule`. To distinguish them in `LogBookModule`, we use the import alias, as shown in Listing 6-15.

Listing 6-15: log-book.module.ts (Exercise-06-04)

```
import {NgModule} from '@angular/core';
import {BrowserModule} from '@angular/platform-browser';

import {LogBookComponent} from './log-book.component';
import {UtilityModule} from '../sharedutil/utility.module';
import {UtilityModule as AcmeUtils} from '../acmeutils/utility.module';


@NgModule({
  imports: [
    BrowserModule,
    UtilityModule,
    AcmeUtils
  ],
  declarations: [LogBookComponent],
  exports: [LogBookComponent]
})
export class LogBookModule { }
```

The "**UtilityModule as AcmeUtils**" expression creates an alias for the identifier of AcmeFactory's **UtilityModule**, so we will not mix it up with GigaCorp's module in the imports section of **@NgModule()**.

NOTE: In the **Exercise-06-04** *folder, you can find the rest of the source code not shown in this section.*

When you run the app, you can check that both directives work. Moving the mouse over the log book UI part triggers both **HighlighterDirective** and **BorderDirective** (Figure 6-8).

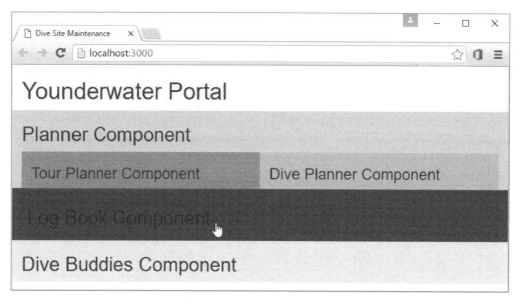

Figure 6-8: The shared modules are added to the wireframe

A Few Comments about Modules

We cannot sign explicitly that a module is a feature module or a shared module. These concepts are only stereotypes and rigid terms. We import feature modules in exactly one module—either in the application root module or another feature module—, while we import shared modules in every other module we need to use them.

Observe, modules are logical containers for objects that need to expose themselves to the Angular framework. These objects either have metadata—such as components, directives, pipes, services—, or are referenced in metadata—as you will learn later in this chapter.

You do not need to put unexposed objects into modules. For example, if you have a frequently used set of files with data structures—such as a class that describes a dive log entry—, those should not be put into a shared module. Just import them into the other source files where they are utilized.

There is one type of object that needs more care than the others: with services, you need to be more circumspect. Soon, you will learn why.

Understanding Services and Dependency Injection

Angular has a crucial concept, the *service*. Anything can be a service that has a narrow, well-defined functionality we intend to use in any other parts of the application, including components, directives, pipes, utility modules, other services, and so on. Evidently, services do not define their user interfaces; they are often utilized by other objects that expose UI. Here are a few examples of potential services:

When you create an e-commerce application, a shopping cart is a good candidate for being a service. Many apps are data-driven and use backend information that rarely changes over time. This kind of apps often works with several simple dictionaries that you render on the UI as drop-down lists, list boxes or another kind of selection controls. From the app's functional point of view, it is enough to read these dictionaries only once—even with the "lazy" pattern—, store them in a cache, and then later to use cached instances, sparing the further and unnecessary communication with the backend.

Services are also good candidates to encapsulate reusable code that does not belong to any other Angular object, but is utilized by many of them. In *Chapter 3*, you created a service, **SiteManagementService**, which queried and persisted dive site information. Other components used a singleton instance of this service.

In Angular, most services are represented by classes, but values and functions can be used as services, too.

We never have to instantiate the classes that represent components, directives, or pipes: Angular undertakes this task for an excellent reason—the framework is the owner of these objects and insist on managing their life cycle.

The objects to be created depend on other instances. Most of these dependencies are services. For example, in Chapter 3, four components, **SiteListComponent**, **AdSiteComponent**, **EditSiteComponent**, and **DeleteSiteComponent** depended on **SiteManagementService**. Angular passes the dependencies through the constructor of objects with its *dependency injection* mechanism. For example, the components mentioned above had a constructor like this to allow the framework to inject a **SiteManagementService** and a **Router** instance:

```
export class SiteListComponent {
  sites: DiveSite[];

  constructor(
    private siteService: SiteManagementService,
    private router: Router
  ) {
    // ...
  }
  // ...
}
```

Angular uses the types of constructor arguments to find out which dependencies to create. The framework creates the requested instances, invokes the constructor and passes the previously created dependent instances.

> *NOTE: You are not entirely tied to the types to define what kind of dependency should be injected for a particular argument. Later in this chapter, you will learn another method.*

Injectors, Providers, and Tokens

When the framework needs to obtain an object (a service) to inject into a constructor, it asks an *injector* to get that object. The injector may respond with success—it retrieves the requested object—or may report a failure—it was unable to get the dependency. The framework manages not only one but more injectors, and it has a fallback strategy to ask other injectors if querying one would fail.

The injector keeps a list of *providers*. A provider is a kind of recipe to get a particular type of dependency. If the injector has a provider for the requested dependency, then it tells the provider to retrieve the object it manages. Providers use a *token* to identify the requested service. With other words, the injector holds a map where tokens are the keys and providers are responsible for retrieving the values.

You may ask why we need the providers. Would not it be simpler if the injector would hold key and value pairs? When we needed a dependency, it would resolve its key (token) and retrieve the value. Well, it sounds much simpler!

Unfortunately, this scenario is far away from most real-life situations. Sometimes, we need to pass a singleton dependency instance to all requesters, while in other cases, we might have to create separate instances for every query. In complicated scenarios, we need to create the same instance for a cluster of components, and another one for a different cluster. Providers offer this kind of flexibility.

To configure the dependency injection mechanism, we need to register providers with injectors. Applications, modules, components, and directives all have injectors, and we can register providers with any of them.

Resolving Dependencies

Let's understand how Angular dependency injection works! This mechanism can be summarized in a few points:

1. When the framework is about to create a new Angular object, it checks the object's constructor for requested dependencies. If there is any, first those dependencies are obtained. The engine gets them by applying this mechanism recursively and traverses the entire dependency chain. When the framework has collected all of them, it invokes the constructor with passing the requested dependencies as arguments.

2. During this search, Angular walks through a chain of injectors and queries them whether they can resolve a particular dependency. This traversal goes on unless the requested object is obtained, or the entire chain is visited with no success.

3. At any phase of this check, the injector gets a token that represents the dependency. With that token, the injector goes through the list of its registered providers to check that any of them accepts the token.

4. If such a provider is found, the injector asks the provider to retrieve the dependency object and then returns the object while reporting success. Otherwise, the injector signs the failure of resolution.

There are two more things to clarify so that you can get acquainted all details of the mechanism:

#1: What is the chain of injectors?

#2: How do providers retrieve the requested object?

Let's find the answers to these questions!

#1. While an app is running, its modules and component indirectly build up a tree of injectors represented by modules, components, directives—and other objects. Figure 6-9 depicts such a tree.

When Angular bootstraps a module, it creates a *platform injector*—as a result of importing `BrowserModule`—, and an injector for the application root modules (`AppModule` in the figure). When a module is loaded, its components, directives, and imported modules also initialize injectors. As a component is rendered, it may create nested child components, which also have injectors.

Thus, while the app runs, the modules, components, and other objects form a dynamically established tree, where each node has an injector. It does not mean that every node in this tree has its dedicated injector—several nodes may share the same—, but each of them points to one.

When a component in this hierarchy needs a dependency to be resolved, the framework traverses through the injectors associated with the tree nodes starting from the component and going up to the platform injector. For example, if an `AppChild` component requests a dependent object, the framework visits `AppChild`, `AppComponent`, `AppModule`, and the platform injector, in this very order.

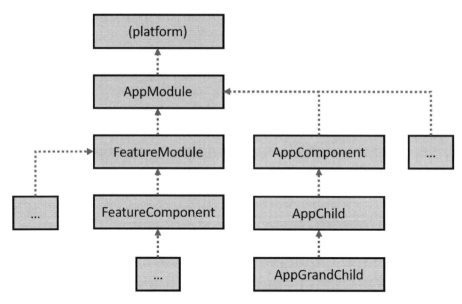

Figure 5-9: The injector chain

#2. As I mentioned earlier, a provider is like a recipe. When the framework asks the provider to get an object for the token presented, the recipe gives clear instructions to obtain that object. Angular has several "predefined recipes"—and savvy Angular developers even can create custom ones.

One of the recipes—**useClass**, as soon you will learn why it has this name—has one ingredient, **className**, and works this way:

"If you already store a previously created dependency object, retrieve that one. Otherwise, take **className**, ask the dependency injection mechanism to create an instance of it (obviously with resolving the requested dependencies beforehand), store the freshly created object, and retrieve it."

According to the description, the **useClass** recipe uses lazy instantiation, and a particular provider always retrieves the same object instance.

Now, you have learned all essential things to understand how Angular injects dependencies. Let's see this mechanism in action.

Dependency Injection in Action

In this section, we are going through several exercises that implement a very basic diving gear checklist app. The UI of the app is simple; it contains toggle buttons to check or uncheck that a particular diving gear is packed. Figure 6-10 shows a snapshot of the simplest version of this app.

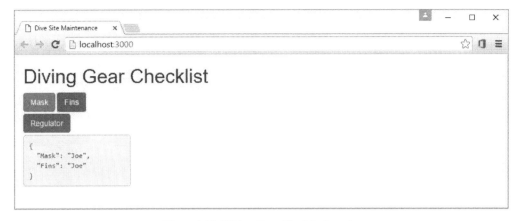

Figure 6-10: Diving Gear Checklist in action

A centerpiece of the app is **InventoryService**. We will modify it only slightly. As we go through the exercises, we change the way this service is injected into application components.

A Few Core Elements of the Sample

As Listing 6-16 shows, the code of **InventoryService** is straightforward—though it bears some unusual things.

Listing 6-16: inventory.service.ts (Exercise-06-05)

```
import {Injectable} from '@angular/core';

@Injectable()
export class InventoryService {
  private _items: {[key: string]: string} = {};

  getItems() {
    return this._items;
  }

  hasItem(key: string): boolean {
    return !!this._items[key]
  }

  toggle(key: string, owner: string) {
    if (this._items[key]) {
      delete this._items[key];
    }
    else {
      this._items[key] = owner;
    }
  }
}
```

Angular needs metadata so that it can automatically manage the lifecycle of objects. Thus we decorate services with the **@Injectable()** decoration—for this sole purpose.

We store the packed items in the **_items** member. The strange "**{[key: string]: string}**" TypeScript notation means that **_items** is an object with string keys (property names) and string values. Using the service, we can retrieve the packed items, check if a particular item is packed, and toggle the state of an item (from packed to unpacked and vice versa). The **toggle()** method accepts an **owner** property that we need for demonstration purposes.

The toggle buttons in Figure 6-10 are implemented by **ItemComponent**, as Listing 6-17 shows. This component builds on **InventoryService**; the button invokes the service to store the packed items.

Listing 6-17: item.component.ts (Exercise-06-05)

```
import {Component} from '@angular/core';
import {InventoryService} from './inventory.service';

@Component({
  selector: 'gear-item',
  template: `
    <button class="btn"
      [style.margin-bottom.px]="4"
      [class.btn-success]="selected"
      [class.btn-danger]="!selected"
      (click)="toggle()">
      {{name}}
    </button>
  `,
  inputs: ['name', 'owner']
})
export class ItemComponent {
  name: string;
  owner: string;
  selected = false;

  constructor(private inventory: InventoryService) {
  }

  toggle() {
    this.inventory.toggle(this.name, this.owner);
    this.selected = this.inventory.hasItem(this.name);
  }
}
```

As the code shows—and this is what you already learned about dependency injection—the dependent service is injected into the constructor of the class.

The app's UI is managed by **AppComponent** that displays diving gear items and a JSON output of the inventory (Listing 6-18).

Listing 6-18: app.component.ts (Exercise-06-05)

```
import {Component} from '@angular/core';
import {InventoryService} from './inventory.service';

@Component({
  selector: 'yw-app',
  template: `
    <div class="container-fluid">
      <h1>Diving Gear Checklist</h1>
      <div class="row">
        <div class="col-sm-3">
          <gear-item *ngFor="let item of items"
            [name]="item"
            [owner]="'Joe'">
          </gear-item>
          <pre>{{inventory.getItems() | json}}</pre>
        </div>
      </div>
    </div>
  `
})
export class AppComponent {
  items = ['Mask', 'Fins', 'Regulator']

  constructor(private inventory: InventoryService) {
  }
}
```

To display the packed items, **AppComponent** utilizes **InventoryService**, too. The application cannot work without instructing the framework about the way **InventoryService** can be injected into the requesting components. We register a provider for the service in the application root module (Listing 6-19).

Listing 6-19: app.component.ts (Exercise-06-05)

```
import {NgModule} from '@angular/core';
import {BrowserModule} from '@angular/platform-browser';

import {AppComponent} from './app.component';
import {ItemComponent} from './item.component';
import {InventoryService} from './inventory.service';

@NgModule({
  imports: [BrowserModule],
  declarations: [
    AppComponent,
    ItemComponent
  ],
```

```
  providers: [InventoryService],
  bootstrap: [AppComponent]
})
export class AppModule { }
```

That is all. The sample app works as we expect. But how does it inject **InventoryService** into **AppComponent** and the instances of **ItemComponent**?

Injecting InventoryService

When we register a provider, we can register either a type or a provider object. When we use a type, Angular automatically interprets it as a provider. In **AppModule**, we used this registration:

```
providers: [InventoryService]
```

It is as if we wrote this:

```
providers: [{provide: InventoryService, useClass: InventoryService}]
```

The **InventoryService** type used in the first registration is translated to a provider object. In this form, **provide** defines the token, **useClass** names the recipe, with one ingredient, the **InventoryService** class. Earlier you already learned that the **useClass** recipe lazy-instantiates a singleton service object—and thus, we expect Angular to inject this singleton instance both into **AppComponent** and **ItemComponent**. Let's see in a few steps how this happens.

When the app bootstraps, it creates **AppModule** with the **InventoryService** provider (Figure 6-11).

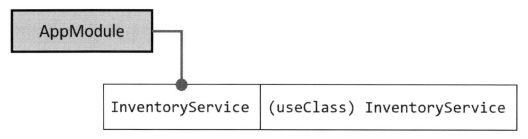

Figure 6-11: AppModule has been just created

Then **AppModule** is about to create **AppComponent**—and first, it needs to have an instance of **InventoryService** so that it can be injected into the component constructor. Before instantiating **AppComponent**, Angular builds a tree node for it (Figure 6-12), copies the **@Component()** metadata to that node, and assigns an injector to the node.

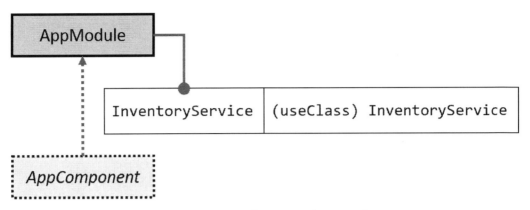

Figure 6-12: AppComponent is being created

Now, the framework needs to obtain an **InventoryService** instance. It gets one with these steps:

First, it checks the **AppComponent** injector for an **InventoryService** provider. Because **AppComponent** did not register any provider, this step fails.

Second, it moves up to **AppModule** and finds the **InventoryService** provider. The provider instantiates a service object—according to the **useClass** recipe—and caches it.

Now, **AppComponent** constructor can be invoked with the injected service instance. Right after **AppComponent** is created, the framework stores its reference in the tree node.

When the view of **AppComponent** is rendered, Angular has to create an **ItemComponent** that has an **InventoryService** dependency, too. The framework starts with a wrapper—and a corresponding injector—for **ItemComponent** (Figure 6-13).

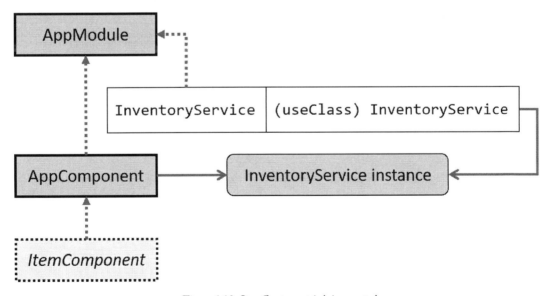

Figure 6-13: ItemComponent is being created

When resolving **InventoryService**, the framework reaches the provider registered with **AppModule**. The provider has already cached a service object, thus now, it retrieves that object. Angular creates **ItemComponent** injected with the same **InventoryService** (Figure 6-14).

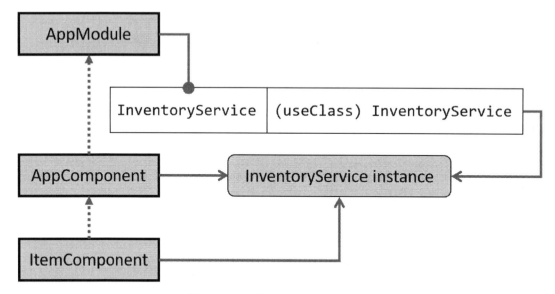

Figure 6-14: All dependencies are resolved

AppComponent has multiple **ItemComponent** children—one for each toggle button. The dependencies of these children are injected into their constructors the same way as you have just learned. The result of the process is that each **ItemComponent** instance utilizes the very same instance of **InventoryService**—the same as injected into **AppComponent**.

> *NOTE: Figure 6-14 shows only one instance of **ItemComponent**—for the sake of simplicity—, but the app creates three of them.*

Working with Multiple Items

Let's make this sample a bit more involved. In the **Exercise-06-06** folder, you find a modified version that introduces a new component, **DiverComponent**, which takes the role of **AppComponent** in the previous example (Listing 6-20).

Listing 6-20: diver.component.ts (Exercise-06-06)

```
import {Component} from '@angular/core';
import {InventoryService} from './inventory.service';

@Component({
  selector: 'diver',
  template: `
```

```
     <div class="col-sm-3">
       <h2>{{name}}</h2>
       <gear-item *ngFor="let item of items"
         [name]="item"
         [owner]="name">
       </gear-item>
       <pre>{{inventory.getItems() | json}}</pre>
     </div>
   `,
   inputs: ['name', 'items']
})
export class DiverComponent {
  name: string;
  items: string[];

  constructor(private inventory: InventoryService) {
  }
}
```

AppComponent renders multiple instances of **DiverComponent** (Listing 6-21).

Listing 6-21: app.component.ts (Exercise-06-06)

```
import {Component} from '@angular/core';

@Component({
  selector: 'yw-app',
  template: `
    <div class="container-fluid">
      <h1>Diving Gear Checklist</h1>
      <div class="row">
        <diver *ngFor="let diver of divers"
          [name]="diver"
          [items]="items">
        </diver>
      </div>
    </div>
  `
})
export class AppComponent {
  divers = ['Joe', 'Cecile', 'Martha', 'Steve']
  items = ['Mask', 'Fins', 'Regulator']
}
```

The sample does not change any of **AppModule**, **ItemComponent**, or **InventoryService**. When you run the app, you can experience that something strange. It seems that divers do not have their dedicated inventories but a mingled one (Figure 6-15).

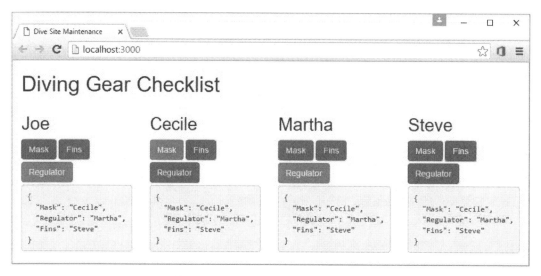

Figure 6-15: Divers with mingled inventories

The reason is pretty clear. Because we had not changed **AppModule**, all **DiverComponent** objects were injected with the same **InventoryService** instance. To resolve this issue and allow each diver to have a dedicated inventory, we need to register the service provider with **DiverComponent** and remove the registration from **AppModule**. The modified sample in **Exercise-06-07** does this, as shown in Listing 6-22.

Listing 6-22: diver.component.ts (Exercise-06-07)

```
import {Component} from '@angular/core';
import {InventoryService} from './inventory.service';

@Component({
  selector: 'diver',
  // ...
  providers: [InventoryService]
})
export class DiverComponent {
  name: string;
  items: string[];

  constructor(private inventory: InventoryService) {
  }
}
```

Now, the app works the way you expect (Figure 6-16).

Figure 6-16: The fixed sample

Let's see how this modification changed dependency injection.

Instead of **AppModule**, we registered **InventoryService** with **DiverComponent**. Thus, when a new instance of **DiverComponent** is about to be created, the framework uses the provider assigned to **DiverComponent** (Figure 6-17).

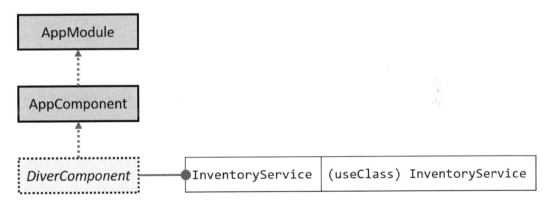

Figure 6-17: DiverComponent is being created

DiverComponent has **ItemComponent** children. When the first is created, it is injected with the same **InventoryService** instance as held by its **DiverComponent** parent (Figure 6-18). Obviously, the other **ItemComponent** children will have the very same service object instance.

The app creates four instances of **DiverComponent** and each has three **ItemComponent** children. First, we may think that all of these objects will be injected with a singleton **InventoryService** instance, but it does not happen. Each **DiverComponent** has its dedicated service object instance.

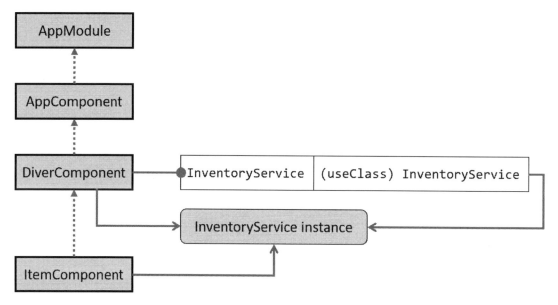

Figure 6-18: ItemComponent created

The key to understanding why it happens this way is the method of component creation. As you have already learned, Angular creates a tree node for the component *before* the component itself is created. The `@Component()` metadata, including the registered providers, is copied to this node. The search for the provider also starts with this node.

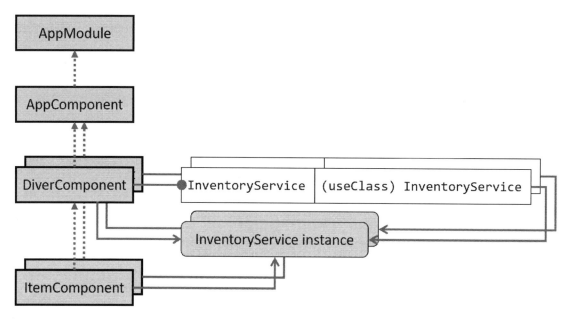

Figure 6-19: All components created

When Angular is about to create the second **DiverComponent** instance, it finds an **InventoryService** provider entry right in the injector associated with the newly created node. Because this provider has not cached a service object yet, it creates and caches a new one (Figure 6-19). When the **ItemComponent** children are created, they will use the very same provider's cached service instance.

Using Service Providers

In the previous samples, you learned the gist of the dependency injection mechanism. So far, you met with only one type of provider, **useClass**, which caches a lazy-instantiated object. In this section, you get acquainted with other provider behaviors.

Injecting a New Service Implementation

With the **useClass** provider, we can inject another class instead of the original service class—for example, a new modified or enhanced version. Let's assume that we create an advanced **InventoryService** that allows tracing the **toggle()** operation (Listing 6-23).

Listing 6-23: inventory.ts (Exercise-06-08)

```
import {Injectable} from '@angular/core';

@Injectable()
export class InventoryService {
  // ...
  toggle(key: string, owner: string) {
    if (this._items[key]) {
      delete this._items[key];
    }
    else {
      this._items[key] = this._defaultOwner || owner;
    }
  }
}

@Injectable()
export class TracedInventoryService {
  // ...
  toggle(key: string, owner: string) {
    if (this._items[key]) {
      delete this._items[key];
      console.log(`Removed ==> ${key}: ${owner}`);
    }
```

```
    else {
      this._items[key] = this._defaultOwner || owner;
      console.log(`Added ==> ${key}: ${owner}`);
    }
  }
}
```

We can quickly inject **TracedInventoryService** into the components which utilize **InventoryService**, as shown in Listing 6-24.

Listing 6-24: diver.component.ts (Exercise-06-08)

```
...
@Component({
  selector: 'diver',
  // ...
  inputs: ['name', 'items'],
  providers: [{
    provide: InventoryService,
    useClass: TracedInventoryService
  }]
})
export class DiverComponent {
  // ...
}
```

The highlighted declaration tells the framework that whenever an **InventoryService** token is presented to the dependency resolver, the provider should create a cached instance of **TracedInventoryService**.

When you run the app, the console output proves that Angular applies the traced version of the service (Figure 6-20).

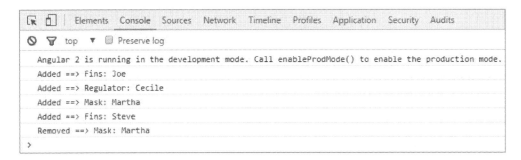

Figure 6-20: TracedInventoryService in action

In Listing 6-23, you can observe that **InventoryService** and **TracedInventoryService** do not have a common ancestor—they are two independent classes. What they have in common is their shape: both have the same methods.

In compiled curly-brace languages such as C, C++, Java, C#, and others, we could use an interface to declare the behavior of these service classes.

We can use the same technique in TypeScript, as TypeScript also supports interfaces (Listing 6-25).

Listing 6-25: inventory.service.ts (Exercise-06-09)

```
import {Injectable} from '@angular/core';

export interface InventoryContract {
  getItems(): void;
  hasItem(key: string): boolean;
  toggle(key: string, owner: string): void;
}

@Injectable()
export class InventoryService implements InventoryContract {
  // ...
}

@Injectable()
export class TracedInventoryService implements InventoryContract {
  // ...
}
```

Changing all references of **InventoryService** to **InventoryContract** and using the interface as the dependency token (Listing 6-26) seems to be an excellent idea.

Listing 6-26: diver.component.ts (Exercise-06-09)

```
...
@Component({
  selector: 'diver',
  // ...
  providers: [{
    provide: InventoryContract,
    useClass: TracedInventoryService
  }]
})
export class DiverComponent {
  name: string;
  items: string[];

  constructor(private inventory: InventoryContract) {
  }
}
```

Unfortunately, it does not work. We cannot use **InventoryContract** as dependency token. JavaScript does not have an explicit construct for interfaces. **InventoryContract** is a name that exists only during the TypeScript compilation—used for semantical checks. In the JavaScript output, this name does not exist at all, so Angular cannot utilize it during run time.

> **HINT:** *Try running the sample in the* **Exercise-06-09** *folder and look at the error messages. It is worth to take a look at the* **inventory.component.js** *file to check that the* **InventoryContract** *name has gone.*

When a Service Depends On Another Service

Having two **InventoryService** classes—one with tracing, one without it—does not help application maintainability. It would be a smarter solution to have a single service class where tracing can be turned on and off. Instead of building switch-flags into the code, we can inject an optional tracing service into **InventoryService** as shown in Listing 6-27 and Listing 6-28.

Listing 6-27: trace.service.ts (Exercise-06-10)

```
import {Injectable} from '@angular/core';

@Injectable()
export class TraceService {
  trace(message: string) {
    console.log(message);
  }
}
```

Listing 6-28: trace.service.ts (Exercise-06-10)

```
import {Injectable, Optional} from '@angular/core';
import {TraceService} from './trace.service';

@Injectable()
export class InventoryService {
  private _items: { [key: string]: string } = {};
  private _defaultOwner: string;

  constructor(@Optional() private tracer: TraceService) { }

  getItems() {
    return this._items;
  }

  hasItem(key: string): boolean {
    return !!this._items[key]
  }

  toggle(key: string, owner: string) {
    if (this._items[key]) {
```

```
      delete this._items[key];
      if (this.tracer) {
        this.tracer.trace(`Removed ==> ${key}: ${owner}`);
      }
    }
    else {
      this._items[key] = this._defaultOwner || owner;
      if (this.tracer) {
        this.tracer.trace(`Added ==> ${key}: ${owner}`);
      }
    }
  }
}
```

The **@Optional()** annotation tells the framework that it must allow the creation of the **InventoryService** class even if the dependency resolution of **TraceService** fails. Angular injects a null value when it cannot resolve an optional dependency.

In the listing, you can see that the code is prepared to handle the null value of **tracer**.

The app needs separate inventory objects for each **DiverComponent** and a singleton **TraceService** instance for the entire app. Accordingly, we register a provider for **InventoryService** in **DiverComponent**:

```
@Component({
  selector: 'diver',
  //
  providers: [InventoryService]
})
export class DiverComponent {
  // ...
}
```

The best place for the **TraceService** registration is **AppModule**:

```
@NgModule({
  imports: [BrowserModule],
  // ...
  providers: [TraceService],
  bootstrap: [AppComponent]
})
export class AppModule { }
```

The app behaves the same way as its previous version. When you run, the app will display similar output you have already seen in Figure 6-20.

When the first **DiverComponent** instance was created, Angular carried out these steps:

#1: The framework recognized that **DiverComponent** needed an **InventoryService** dependency.

#2: **InventoryService** required a **TraceService** dependency.

#3: Angular resolved **TraceService** with the provider registered in **AppModule**.

#4: The framework resolved **InventoryService** with the provider registered in **DiverComponent**. When constructing **InventoryService**, the service object obtained in step #3 was injected.

#5: The framework created **DiverComponent** with the inventory object acquired in the previous step.

> **HINT**: *Remove the provider registration from* **AppModule**, *and check that the app still works. This time, it does not trace using the inventory operations.*

Using Service Aliases

As time goes on, we improve **TraceService**. The advanced version adds extra information to the messages. Listing 6-29 shows that we count the instances of the service created, and display that count.

Listing 6-29: trace.service.ts (Exercise-06-11)

```
import {Injectable} from '@angular/core';

@Injectable()
export class TraceService {
  trace(message: string) {
    console.log(message);
  }
}

@Injectable()
export class AdvancedTraceService {
  static count = 0;
  instance: number;

  constructor() {
    this.instance = ++AdvancedTraceService.count;
  }

  trace(message: string) {
    console.log(`Instance: ${this.instance} - ${message}`);
  }
}
```

Assume that our app uses **TraceService** heavily. In the sample, we inject it only to **AppComponent** and **InventoryService**, but let's call it "heavy" usage. Fortunately, we need to change only the **AppModule** provider registration to upgrade **TraceService** to **AdvancedTraceService**:

```
@NgModule({
  //
  providers: [
    {provide: TraceService, useClass: AdvancedTraceService}
  ],
  bootstrap: [AppComponent]
})
export class AppModule { }
```

Figure 6-21 shows that now the app displays the instance identifier in each log message.

Figure 6-21: The output of the advanced tracer

AdvancedTraceService works so fine that we decide to change all **TraceService** references gradually in the whole application. Our strategy is that we replace the references in the modified source code in every development sprint. This approach leads to duality: for a particular period we have both **TraceService** and **AdvancedTraceService** references, and thus we need to use both tokens. No problem, as shown in Listing 6-30, we can register two providers.

Listing 6-30: app.module.ts (Exercise-06-12)

```
...
@NgModule({
  imports: [BrowserModule],
  // ...
  providers: [
    AdvancedTraceService,
    {provide: TraceService, useClass: AdvancedTraceService}
  ],
  bootstrap: [AppComponent]
})
export class AppModule { }
```

This method makes our components work seamlessly independently of whether they use **TraceService** or **AdvancedTraceService**. In **Exercise-06-12**, **AppComponent** utilizes the old version of tracing, while **InventoryService** goes on with the new service. Nonetheless, there is a little issue with this dual registration, as indicated in Figure 6-22. The app uses two instances of **AdvancedTraceService**.

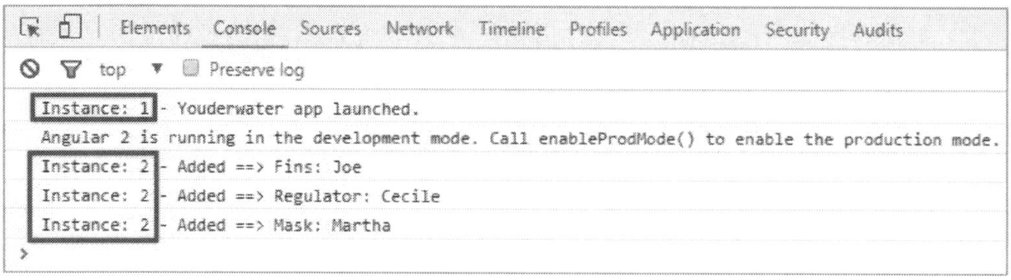

Figure 6-22: The app uses two tracer instances

From the declaration of providers, the reason is clear. Both providers work with cached service instances, so each of them creates and hold one.

We can fix this issues with the **useExisting** provider, as shown in Listing 6-31.

Listing 6-31: app.module.ts (Exercise-06-13)

```
...
@NgModule({
  imports: [BrowserModule],
  // ...
  providers: [
    AdvancedTraceService,
    {provide: TraceService, useExisting: AdvancedTraceService}
  ],
  bootstrap: [AppComponent]
})
export class AppModule { }
```

In this scenario, **useExisting** returns an object as if the **AdvancedTraceService** token was used. In other words, we can say that this provider redirects the dependency resolution to another token—as named in the **useExisting** property of the registration object.

This approach takes care that a singleton **AdvancedTraceService** object is created in the app, independently which of the two tokens we use.

*HINT: Run the **Exercise-06-13** sample to check that only a singleton instance of the service is used.*

Registering a Service Instance

There are several situations when we intend to use a singleton instance of the service and instead of the framework, we intend to create that service object. The sample in the **Exercise-06-14** folder demonstrates this situation. As Listing 6-32 shows, **AdvancedTraceService** has an **info** property to set an additional part of the log message.

Listing 6-32: trace.service.ts (Exercise-06-14)

```
import {Injectable} from '@angular/core';

@Injectable()
export class TraceService {
  trace(message: string) {
    console.log(message);
  }
}

@Injectable()
export class AdvancedTraceService {
  info: string;

  trace(message: string) {
    console.log(`Info: ${this.info} - ${message}`);
  }
}
```

Listing 6-33: app.module.ts (Exercise-06-14)

```
import {NgModule} from '@angular/core';
import {BrowserModule} from '@angular/platform-browser';

import {AppComponent} from './app.component';
import {ItemComponent} from './item.component';
import {DiverComponent} from './diver.component';
import {TraceService, AdvancedTraceService} from './trace.service';

let tracer = new AdvancedTraceService();
tracer.info = "V0.12";

@NgModule({
  imports: [BrowserModule],
  declarations: [
    AppComponent,
    ItemComponent,
    DiverComponent
  ],
  providers: [
    {provide: AdvancedTraceService, useValue: tracer},
    {provide: TraceService, useExisting: AdvancedTraceService}
  ],
  bootstrap: [AppComponent]
})
export class AppModule { }
```

If we rely on the framework to create an instance of **AdvancedTraceService**, we can be sure that **info** remains undefined. To set **info** to a useful value, we need to create an object instance with **info** set and register a provider for that very instance (Listing 6-33).

The **useValue** provider—as its name suggests—registers a concrete pre-created value with the injector.

When you run the sample, the console output shows that log entries contain the "V0.12" information we set in the listing (Figure 6-23).

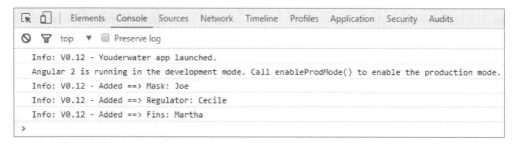

Figure 6-23: The extra information is displayed

Using a Service Factory

Sometimes we need a tight control on obtaining dependency objects. We may use specialized caching, switching object type at run time, using a pool of service objects, and so on.

Angular allows us to register a provider with a factory. The sample in **Exercise-06-15** uses this feature to create instances of **InventoryService** with an initial set of items that divers have in their package by default. Such articles are a torch and a buoyancy control device (BCD), and these are brought by a pal named Dave (Figure 6-24).

Listing 6-34 shows the declaration of a function, **invFactory()**. This function does not retrieve a service instance. Instead, it returns a function—the *factory function*—that can be invoked to obtain the object.

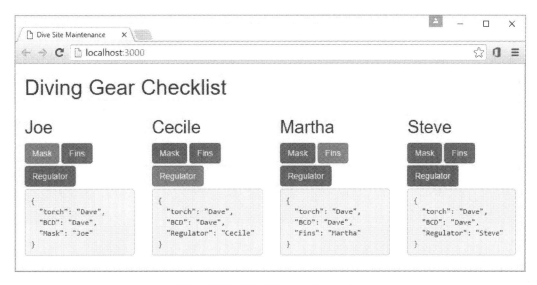

Figure 6-24: Default items in the inventory

Listing 6-34: diver.component.ts (Exercise-06-15)

```
...
@Component({
  selector: 'diver',
  // ...
  inputs: ['name', 'items'],
  providers: [{
    provide: InventoryService,
    useFactory: invFactory(['torch', 'BCD'], 'Dave'),
    deps: [AdvancedTraceService]
  }]
})
export class DiverComponent {
  name: string;
  items: string[];

  constructor(private inventory: InventoryService) {
  }
}

function invFactory(init: string[], owner: string) {
  return (tracer: AdvancedTraceService) => {
    var invSrv = new InventoryService(tracer);
    for (let i = 0; i < init.length; i++) {
      invSrv.toggle(init[i], owner);
    }
    return invSrv;
  }
}
```

The framework invokes the returned function and passes the resolved dependency objects to it. The listing clearly demonstrates this operation. To instantiate an **InventoryService**, we need an **AdvancedTraceService**. This dependency is passed in the **tracer** argument of the factory function.

Take a look at the provider registration. The **useFactory** property is set to the result of the **invFactory()** call; the real factory function is the one returned by this call. Besides **useFactory**, we register additional providers with the **deps** property. Because we need to resolve the dependencies of **InventoryService**, we register the providers used for this purpose with **deps**.

> *HINT: You can reuse a factory if you assign the returned function to a variable and pass this initialized variable to **useFactory**.*

Using An OpaqueToken

So far we applied only class names as tokens. Angular accepts any object or value as a provider token, and we can resolve dependencies to numbers, Booleans, strings, or an arbitrary data structure—and not only to class instances. If we use a simple value—such as a string—as a token, we may have colliding token names. For example, two different modules may use the "config" string accidentally as tokens to different objects.

To avoid the issues that might come from these collisions, Angular provides an **OpaqueToken** type to create unique tokens. We store such a token in an exported variable so that we can utilize it in multiple locations to register providers (Listing 6-35).

Listing 6-35: version-token.ts (Exercise-06-16)

```
import {OpaqueToken} from '@angular/core';

export let VERSION = new OpaqueToken('version');
```

When we want to identify a dependency with an **OpaqueToken**—or with anything different from a class name—we need to sign this intention with the **@Inject()** decorator, as shown in Listing 6-36.

Listing 6-36: trace-service.ts (Exercise-06-16)

```
import {Injectable, Inject} from '@angular/core';
import {VERSION} from './version-token';

// ...

@Injectable()
export class AdvancedTraceService {
  info: string;
```

```
  constructor(@Inject(VERSION) private version: string) {}

  trace(message: string) {
    console.log(`Info: ${this.version} - ${message}`);
  }
}
```

When the framework is about to create an **AdvancedTraceService**, it checks the **VERSION** token to resolve the string for the **version** argument. We can register the **useValue** provider to declare the value of **VERSION** to inject into the service (Listing 6-37).

Listing 6-37: app.module.ts (Exercise-06-16)

```
...
@NgModule({
  imports: [BrowserModule],
  // ...
  providers: [
    AdvancedTraceService,
    {provide: VERSION, useValue: 'v0.23'},
    {provide: TraceService, useExisting: AdvancedTraceService}
  ],
  bootstrap: [AppComponent]
})
export class AppModule { }
```

Summary

An Angular module is a container that consolidates pieces of an application or a library into a cohesive block of functionality. It is our design decision what we encapsulate into a module.

Each application has a root module that bootstraps the app. We can group the app's constituent parts into—even hierarchical—feature modules. Modules can be shared and imported by other modules within the app. We can ship individual modules to be used as third-party modules in other systems, too.

Angular has a crucial concept, the *service*. Anything can be a service that has a narrow, well-defined functionality we intend to use in any other parts of the application, including components, directives, pipes, utility modules, other services, and so on.

Services are injected into Angular objects with the dependency injection mechanism. The framework provides a simple way for basic dependency injection scenarios as well as advanced methods for involved cases.

Chapter 7: Working with Components

WHAT YOU WILL LEARN IN THIS CHAPTER

Understanding how components manage the UI

Getting acquainted with the lifecycle events of components

Creating component hierarchies

Understanding the ways you can communicate between parent and child components

Using services for communication between components

Distinguishing a component's content children from its view children

Creating components that transpose their content children to view children

No doubt, the concept of *components* is one of the pivotal things in Angular. In every sample so far you met many components, and now you know the fundamentals of creating and using them. In this chapter, you will understand how they work and what great stunts you can do with them.

Understanding Components

A component represents a part of the UI and controls its behavior. The logic behind the component determines the data displayed in the corresponding UI and responds to user actions. Angular calls the particular UI the *view* of the component.

The logic is represented by a class that defines its API. This API—it is accessible from the view— declares properties and methods that the view can utilize. Whenever the user carries out an action in the view, the view can execute statement—setting properties or invoking methods. When the logic changes some information that should be displayed in the view, it updates properties. The Angular change detection mechanism observes these modifications and refreshes the view accordingly.

Figure 7-1 summarizes the fundamental Angular concepts that belong to components.

Components never manipulate the DOM directly. Instead of modifying the view imperatively, they do it indirectly through properties and methods. Each component has a template that is a declarative description of how the view should be rendered. The template syntax allows a mechanism, *data binding,* which is utilized by the Angular rendering engine to create the DOM of the view. The framework leverages another kind of bindings, *event binding*, to trigger the execution of component logic when the user (or the system) interacts with the view.

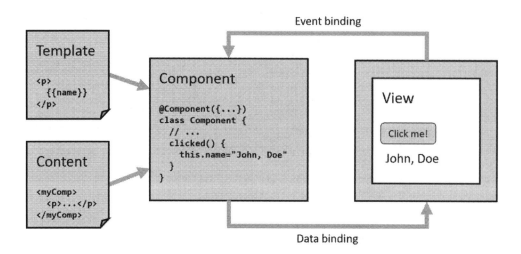

Figure 7-1: The component and related concepts

Components may have content—nested HTML markup within the selector element's opening and closing tag. The component class can access its content and utilize it to modify its view.

Component Metadata

Angular recognizes from the **@Component()** annotation that a class is a component. Without this decoration, the framework could not manage the class because it would not know what to do with it. The **@Component()** annotation accepts a metadata object with about a dozen properties—most of them are optional.

You already met with a few of them—**selector**, **template**, **templateUrl**, **inputs**, **outputs**, **providers**—and, later in this chapter, you will learn about the others, too.

Component Hierarchy

Angular apps utilize multiple components—each managing its dedicated part of the UI. As the patches of UI nest in each other and build up a hierarchy, so do the component instances that represent those parts.

In the **Exercise-07-01** folder, you find a simple application that demonstrates the hierarchy of components. In Figure 7-2, you can observe how their views form a tree.

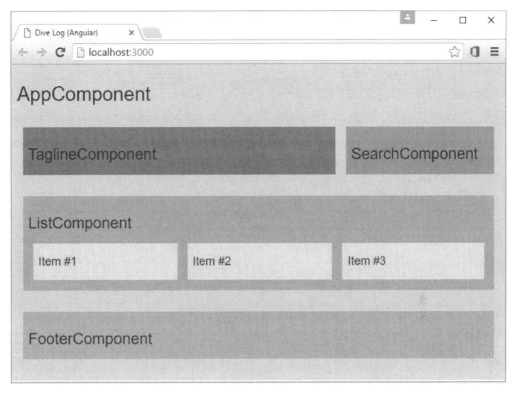

Figure 7-2: Views of component represent a hierachy

As you see, **AppComponent** embeds the others, exactly one instance of its each direct child, **TaglineComponent**, **SearchComponent**, **ListComponent**, and **FooterComponent**, respectively. **ListComponent** also has children—three instances of **ItemComponent**.

Component classes may have a hierarchy—just as types and classes have in OOP languages. The component instances may build up hierarchies even without hard-coded inter-class dependencies, such as inheritance, aggregation, and so on.

When you check the source code of the sample, you can observe that all component classes are entirely independent of each other. Listing 7-1 shows the declaration of **AppComponent** that nest instances from all other classes.

Listing 7-1: app.component.ts (Exercise-07-01)

```
import {Component} from '@angular/core';

@Component({
  selector: 'yw-app',
  template: `
    <div class="container-fluid">
      <h2>AppComponent</h2>
      <div class="container-fluid">
        <div class="row">
```

```
          <div class="col-sm-8">
            <yw-tagline></yw-tagline>
          </div>
          <div class="col-sm-4">
            <yw-search></yw-search>
          </div>
        </div>
        <div class="row">
          <div class="col-sm-12">
            <yw-list></yw-list>
          </div>
        </div>
        <div class="row">
          <div class="col-sm-12">
            <yw-footer></yw-footer>
          </div>
        </div>
      </div>
    </div>
  `,
  styles: [`
    div {
      background-color: #e0e0e0;
      padding: 8px;
    }
  `]
})
export class AppComponent {}
```

When running the app, Angular observes this hierarchy and builds up a tree of instances according to the template. This tree is essential: any node within this tree can communicate with its parent and its children according to the app's logic.

The affected *component instances* form the hierarchy in Figure 7-3.

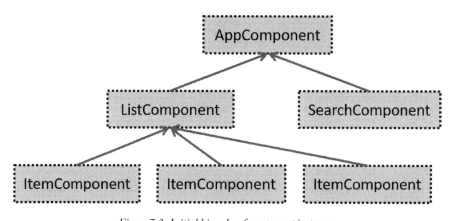

Figure 7-3: Initial hierachy of component instances

Let's assume that we want to refresh the items in the list as the search key changes. When the user types a new search key, **SearchComponent** communicates with its parent, **AppComponent**, and sends a message about the new search key. When **AppComponent** receives this message, it knows that **ListComponent** is affected by the new search key. Thus **AppComponent** notifies **ListComponent** to update itself. **ListComponent** refreshes its children according to the new search text (Figure 7-4).

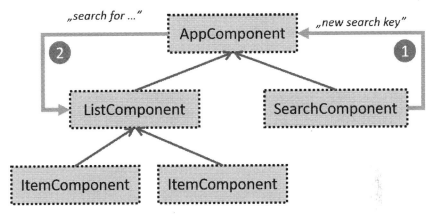

Figure 7-4: Communication within the tree

Component Content

So far we added components to the markup (in **index.html**) with empty or single text content like these:

```
<add-site-view></add-site-view>
...
<yw-app>Loading...</yw.app>
```

Just as the **<ul>** HTML tag may have nested content—a set of **<li>** tags—, we can add content to components, too. For example, we could have a **DashBoardComponent** with content—**<gadget>** items—, like this markup shows:

```
<dashboard-component>
  <gadget type="files" />
  <gadget type="folders" />
  <gadget type="shares" />
</dashboard-component>
```

The component class can access its content and utilize the nested elements to render a particular view that depends on the concrete content. Moreover, when other Angular objects alter the content, the host component can catch these modifications and respond to changes.

Lifecycle Hooks

In the apps, we create component classes, but we never need to instantiate them. Angular owns these components and manages their entire lifecycle. Creates them so that they can be attached to the DOM, follows changes in their properties, and last destroys them when they are detached from the DOM and not needed anymore.

There are situations when we need to know that the component reached a particular milestone in its life, or something—to which we should respond to—has been changed in the state of the component. Angular offers *lifecycle hooks* so that we can catch the moments these important events happen. Component methods with their arguments can represent these hooks. By declaring these lifecycle hook methods, we can act when these special events occur.

In Chapter 3, we already met with the **OnChanges** lifecycle hook. We used it to detect the moment when the **siteId** property of **DeleteSiteComponent** had changed so that we could obtain the site information from a backend service:

```
...
export class DeleteSiteComponent implements OnChanges{
  @Input() siteId: number;
  siteName: string;
  // ...

  constructor(private siteService: SiteManagementService) {
  }

  ngOnChanges() {
    this.siteName = this.siteService
      .getSiteById(this.siteId).name;
  }
  // ...
}
```

We could not have moved the body of **ngOnChanges()** into the constructor because at the moment of the component instantiation the **siteId** input property had not been set. We had to wait for the **OnChanges** event, which signed that Angular had already set data-bound input properties.

OnChanges is only one of the lifecycle hooks. You will learn about others later in this chapter.

> *NOTE: Components are directives. A few lifecycle hooks are available both for directives and components.*

Component Communication

Decomposing your entire applications into smaller components with clear responsibility boundaries help provide better maintainability. Nonetheless, you need to wire up the smaller components into bigger ones and establish the entire application as a set of interacting—communicating—components.

Angular has an excellent toolset to make this interaction straightforward. In this section, you will learn several patterns that you can apply when designing and implementing the communication between your components.

Underwater Token Game Sample

To understand the essential concepts, we are going to implement a simple game, in which scuba divers search for underwater tokens. The game is built up from three components, `AppComponent`, `GameComponent`, and `DiverComponent`. Figure 7-5 shows the UI of the app.

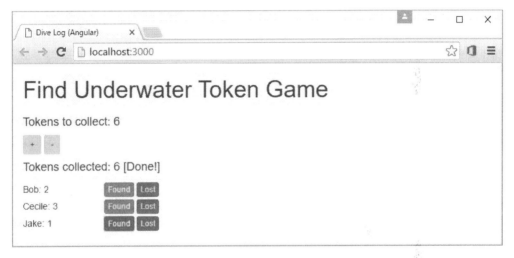

Figure 7-5: The UI of the sample

`AppComponent` is a simple wrapper around the game. `GameController` does the lion's share of the work: it allows setting the target count of tokens to collect, displays the status of the game, and keeps a list of players. A `DiverComponent` instance represents a player who can sign finding or losing a token.

Communicating with Input and Event Bindings

In this game, `GameComponent` and `DiverComponent` cooperate to establish the app's logic. `GameComponent` passes the name of a particular player to the corresponding `DiverComponent` instance. `DiverComponent` notifies `GameComponent` whenever any of the divers finds or loses a token.

Listing 7-2 shows the template of `GameComponent`. It clearly shows the simple logic that changes the count of tokens to be collected and evaluates the current status of the game.

Listing 7-2: game.temlate.html (Exercise-07-02)

```html
<div class="row">
  <div class="col-sm-12">
    <h4>Tokens to collect: {{tokens}}</h4>
    <button class="btn btn-sm"
      [disabled]="tokens > 100"
      (click)="tokens=tokens+1">
      +
    </button>
    <button class="btn btn-sm"
      [disabled]="tokens < 1"
      (click)="tokens=tokens-1">
      -
    </button>
  </div>
</div>
<div class="row">
  <div class="col-sm-12">
    <h4>Tokens collected: {{collected}}
      <span *ngIf="collected >= tokens">
        [Done!]
      </span>
    </h4>
  </div>
</div>
<yw-diver *ngFor="let diver of divers"
  [name]="diver"
  (onTokenFound)="tokenFound($event) ">
</yw-diver>
```

The bottom of the template defines the markup of **DiverComponent**. This piece has a vital role in the cooperation of the two components. To get acquainted what this markup does, take a look at the **GameComponent** class (Listing 7-3).

Listing 7-3: game.component.ts (Exercise-07-02)

```typescript
import {Component} from '@angular/core';

@Component({
  selector: 'yw-game',
  templateUrl: 'app/game.template.html'
})
export class GameComponent {
  tokens = 4;
  collected = 0;
  divers = ["Bob", "Cecile", "Jake"]
```

```
    tokenFound (newTokens: number) {
      this.collected += newTokens;
    }
}
```

GameComponent stores the list of players in the **divers** array. With the **ngFor** directive, the template renders a **DiverComponent** instance for each player in the collection. According to this markup, Angular creates a parent-child relationship between **GameComponent** (parent) and **DiverComponent** (child). Because there are three items in **divers**, the parent will have three children:

```
<yw-diver *ngFor="let diver of divers"
  [name]="diver"
  (onTokenFound)="tokenFound($event) ">
</yw-diver>
```

Observe that **GameController** passes the name of the diver to a child with property binding: **[name]="diver"**. This binding assumes that **DiverComponent** has an *input property*, **name**.

When a diver finds or loses a token, **DiverComponent** raises an event, **onTokenFound**, and sends a number (1 or -1) in the event parameters to sign whether the event is about finding or losing the token. Whenever **GameComponent** gets notified about such an event, it invokes the **tokenFound()** method—with the event parameter passed—to update the number of tokens found.

As you already learned, this kind of binding is event binding. It assumes that **DiverComponent** has a corresponding *output property*, **onTokenFound**.

Listing 7-4 shows how **DiverComponent** declares its input and output properties.

Listing 7-4: diver.component.ts (Exercise-07-02)

```
import {Component} from '@angular/core';
import {Input, Output, EventEmitter} from '@angular/core';

@Component({
  selector: 'yw-diver',
  templateUrl: 'app/diver.template.html'
})
export class DiverComponent {
  @Input() name: string;
  @Output() onTokenFound = new EventEmitter<number>();
  tokensFound = 0;

  found() {
    this.tokensFound++;
    this.onTokenFound.emit(1);
  }
```

```
  lost() {
    this.tokensFound--;
    this.onTokenFound.emit(-1);
  }
}
```

You can add the **@Input()** or **@Output()** decorators to component properties—or as you will see later, to accessors—to mark input and output properties, respectively. Optionally, you can specify attribute names in the decorator arguments, given the name of a property differs from the attribute used. For example, these decorators would use the **diver** and **found** attribute names in the markup:

```
export class DiverComponent {
  @Input('diver') name: string;
  @Output('found') onTokenFound = new EventEmitter<number>();
// ...
}
```

Instead of applying decorators, you can configure input and output properties with the **inputs** and **outputs** metadata:

```
@Component({
  selector: 'yw-diver',
  templateUrl: 'app/diver.template.html',
  inputs: ['name'],
  outputs: ['onTokenFound']
})
export class DiverComponent {
  name: string;
  onTokenFound = new EventEmitter<number>();
  // ...
}
```

Input and output properties are a part of the component's public API. During run time, Angular checks whether a property in the markup is an input property of a child component. If not, it raises an error message. For example, forgetting to add the **@Input()** annotation to **name** would result in an error with this messages: *"Can't bind to 'name' since it isn't a known property of 'yw-diver'"*.

Using Tightly-Coupled Components

Observe, in the communication pattern we used in the previous sample, components were loosely-coupled. We did not use an explicit reference to the other component type either in **GameComponent** or **DiverComponent**. The logic of **GameComponent** parent did not utilize the fact that its children are **DiverComponent** instances. It would have allowed other component types to raise the **onTokenFound** event, provided the markup had created the appropriate output event binding. The **DiverComponent** could have been the child component of other parents that could have supported its **name** input property and exploited its **onTokenFound** event.

If components are a part of the same logic—and we do not intend to utilize them in other, more generic context—, we may use tightly coupled communication between them. For example, we could pass the parent component's reference to its children.

Input property binding tempts us to supply such a reference through an input property. Take a look at Listing 7-5 and Listing 7-6 that implement this technique. The **getMe()** method of **GameComponent** retrieves a reference to the component itself. In **DiverComponent** we obtain this reference through the **parent** input property and utilize it to invoke the **tokenFound()** method.

Listing 7-5: game.component.ts (Exercise-07-03)

```
import {Component} from '@angular/core';

@Component({
  selector: 'yw-game',
  templateUrl: 'app/game.template.html'
})
export class GameComponent {
  tokens = 4;
  collected = 0;
  divers = ["Bob", "Cecile", "Jake"]

  getMe() {
    return this;
  }

  tokenFound(newTokens: number) {
    this.collected += newTokens;
  }
}
```

Listing 7-6: diver.component.ts (Exercise-07-03)

```
import {Component} from '@angular/core';
import {Input} from '@angular/core';
import {GameComponent} from './game.component';

@Component({
  selector: 'yw-diver',
  templateUrl: 'app/diver.template.html'
})
export class DiverComponent {
  @Input() name: string;
  @Input() parent: GameComponent;
  tokensFound = 0;

  found() {
    this.tokensFound++;
    this.parent.tokenFound(1);
  }
```

```
  lost() {
    this.tokensFound--;
    this.parent.tokenFound(-1);
  }
}
```

Of course, this approach requires a tiny change in the markup that renders **DiverComponents** (Listing 7-7).

Listing 7-7: game.template.html (Exercise-07-03)

```
...
<yw-diver *ngFor="let diver of divers"
  [name]="diver"
  [parent]="getMe()">
</yw-diver>
```

This approach works, but it is not elegant. Angular supports such tightly-coupled situations with its dependency injection mechanism. We can pass a parent reference to its children through the child component's constructor just like we can inject services. We do not need to change the parent component's logic at all.

The sample in the **Exercise-07-04** folder demonstrates this solution. Listing 7-8 shows the slight changes in **DiverComponent**.

Listing 7-8: diver.component.ts (Exercise-07-04)

```
import {Component} from '@angular/core';
import {Input} from '@angular/core';
import {GameComponent} from './game.component';

@Component({
  selector: 'yw-diver',
  templateUrl: 'app/diver.template.html'
})
export class DiverComponent {
  @Input() name: string;

  constructor(private parent: GameComponent) {}

  tokensFound = 0;

  found() {
    this.tokensFound++;
    this.parent.tokenFound(1);
  }

  lost() {
    this.tokensFound--;
    this.parent.tokenFound(-1);
```

```
  }
}
```

This solution makes the **DiverComponent** markup in **GameComponent** even shorter:

```
<yw-diver *ngFor="let diver of divers"
  [name]="diver">
</yw-diver>
```

With the constructor injection approach, we can pass not only a child component's direct parent but any other parent up in the hierarchy. Moreover, we can inject multiple parents—provided we have more than one levels up while we reach the root—each of them in a separate constructor argument.

Communication Through a Service

Sometimes we need communication between unrelated components that are not in a parent-child relationship. Let's assume that we want to follow in a log every event when divers find or lose a token. Later, we would like to use the same log mechanism to add any game events—in whatever components they occur.

We have a **BoardComponent** that can display messages, and want to be sure that other constituent parts of the app can show their messages in the board—independently of where we put **BoardComponent** in the UI.

Components can communicate through a service. We apply this approach so that **GameComponent** and **DiverComponent** can send their messages to **BoardComponent**. Listing 7-9 shows the **MessageBusService** that works as a hub between the communicating parties.

Listing 7-9: message-bus.ts (Exercise-07-05)

```
import {Injectable} from '@angular/core';
import {Subject} from 'rxjs/Subject';

@Injectable()
export class MessageBusService {
  private messageSource = new Subject<string>();

  messageStream = this.messageSource.asObservable();

  sendMessage(message: string) {
    this.messageSource.next(message);
  }
}
```

The class provides the **sendMessage()** method in its API so that components can announce their messages. The code applies the observables pattern and provides a **messageStream** property that can be used to listen to messages that **MessageBusService** receives. Every time the service catches a message through **sendMessage()**, listeners—entities that subscribe to **messageStream**—are be notified.

BoardComponent (Listing 7-10) subscribes to the stream of messages.

Listing 7-10: board.component.ts (Exercise-07-05)

```
import {Component} from '@angular/core';
import {MessageBusService} from './message-bus.service';

@Component({
  selector: 'yw-board',
  templateUrl: 'app/board.template.html'
})
export class BoardComponent {
  messages: string[] = [];

  constructor(private messenger: MessageBusService) {
    messenger.messageStream.subscribe(
      message => { this.messages.push(message); }
    );
  }
}
```

The **messageStream** property of the injected **MessageBusService** is an observable object and provides a **subscribe()** method with a callback function argument. This function is invoked asynchronously every time the service receives a new message.

GameComponent and **DiverComponent** use the familiar pattern to access **MessageBusService**. Listing 7-11 shows the **GameComponent** implementation.

Listing 7-11: game.component.ts (Exercise-07-05)

```
...
@Component({
  selector: 'yw-game',
  templateUrl: 'app/game.template.html'
})
export class GameComponent {
  // ...
  constructor(@Optional() private messenger: MessageBusService) {
  }

  tokenFound(newTokens: number) {
    this.collected += newTokens;
    if (this.messenger) {
      this.messenger.sendMessage(
        `Game ==> Tokens found: ${this.collected}`);
    }
  }
}
```

When you run the app, it displays the messages (Figure 7-6).

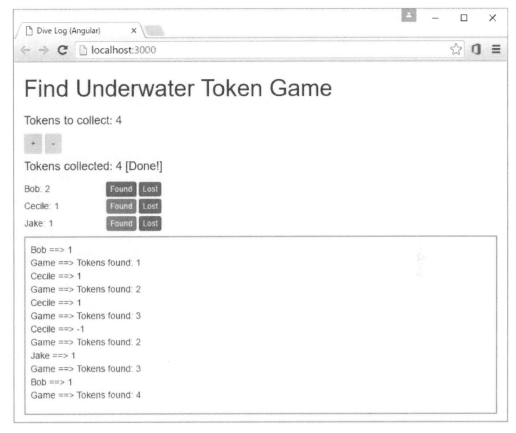

Figure 7-6: Messages went through MessageBusService

> **NOTE:** *Although this sample demonstrated unidirectional communication between unrelated components, we can easily create bidirectional one by setting up another service method with its message stream to represent the other direction.*

Intercepting Property Value Changes

Output event properties allow the receiving party—generally the parent in a child-to-parent communication scenario—to be notified when something changes at the child's side. Occasionally it is useful—or even indispensable—if the children can observe that the parent changes the value of an input property.

Two techniques are helpful in such a situation. First, we can create property setters to catch the event when a property changes. Second, we can implement the **OnChanges** lifecycle hook in the child component.

The **Exercise-07-06** and **Exercise-07-07** folders contain samples that demonstrate these techniques. Both examples catch the event when the **name** input property of **DiverComponent** changes, and they send a message to the board. Listing 7-12 shows the property setter approach.

Listing 7-12: diver.component.ts (Exercise-07-06)

```
...
@Component({
  selector: 'yw-diver',
  templateUrl: 'app/diver.template.html'
})
export class DiverComponent {
  private _name: string;

  @Input() get name() {
    return this._name;
  }

  set name(value: string) {
    this._name = value;
    if (this.messenger) {
      this.messenger.sendMessage(
        `Diver name set ==> ${value}`);
    }
  }

  constructor(
    private parent: GameComponent,
    @Optional() private messenger: MessageBusService) {}
  // ...
}
```

While a property setter can watch only the change of a particular property, the **OnChanges** lifecycle hook allows catching the modification of any input property. Within the hook method, you can check which property or properties are altered. Listing 7-13 demonstrates this approach.

Listing 7-13: diver.component.ts (Exercise-07-07)

```
import {Component, Optional} from '@angular/core';
import {Input, OnChanges} from '@angular/core';
import {GameComponent} from './game.component';
import {MessageBusService} from './message-bus.service';

@Component({
  selector: 'yw-diver',
  templateUrl: 'app/diver.template.html'
})
export class DiverComponent implements OnChanges {
  @Input() name: string;
```

```
constructor(
  private parent: GameComponent,
  @Optional() private messenger: MessageBusService) {}
// ...

ngOnChanges(changes) {
  var nameChange = changes['name'];
  if (nameChange && this.messenger) {
    this.messenger.sendMessage(
      `Diver name set ==> ${nameChange.currentValue}`);
  }
}
}
```

The `OnChanges` lifecycle hook declares a single method, `ngOnChanges()`. Whenever Angular change detection observes that an input property's value is modified, it invokes the hook method. A single run of detection cycle may catch more property changes. These are collected and passed to `ngOnChanges()` in its argument. The parameter object contains a key for each changed property. The value that belongs to a particular key is a `SimpleChange` object instance. Through this object, you can access the `previousValue` and the `currentValue` properties and test the result of `isFirtsChange()`.

Figure 7-7 shows the messages logged by `ngOnChanges()`.

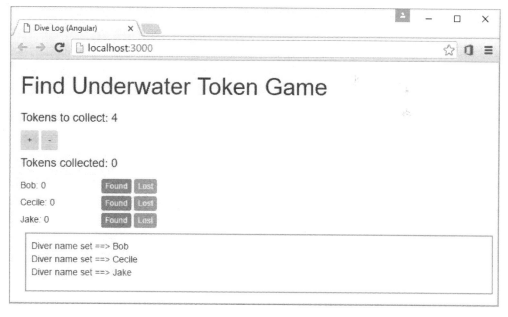

Figure 7-7: The messages logged by ngOnChanges()

Using Component Content

We can add elements between the opening and closing tags of our component's host element. These are called *content children*, and we can utilize them to create the view of the component.

In this section, we take a look at several samples that work with the content.

View Children and Content Children

Let's assume that we have two components. One, **AcmeComponent** is associated with a selector, "acme", and another, **SuperbComponent** with the "superb" selector. In the main page of the app, we use this markup:

```
<!-- index.html -->
<h1>My Acme Solution</h1>
<acme>
    <h2>I am a content child of acme</h2>
    <p>I am, too</p>
</acme>
```

Here, the **<h2>** and **<p>** tags are content children of **AcmeComponent**, because they are located between the opening and closing **<acme>** tags. Similarly, **<h3>** and **<p>** are content children of **AcmeComponent**, too, in the template of **SuperbComponent**:

```
...
@Component({
  selector: 'superb',
  template: `
    <h1>My SuperbComponent</h1>
    <acme>
        <h3>I am a content child of acme</h3>
        <p>I am, too</p>
    </acme>
  `
})
export class SuperbComponent {
}
```

The elements which are located inside the template of a component are called *view children*. Thus, in **SuperbComponent**'s declaration, **<h1>**, **<acme>**, **<h3>**, and **<p>** are all view children.

By default, Angular does not render content children. It relies on the host component to manage its children and render them according to the component's intention.

The sample in the **Exercise-07-08** folder demonstrates this fact. It's **main.ts** file is modified so that the root application module is bootstrapped about two seconds after the page is loaded:

```
import {platformBrowserDynamic} from '@angular/platform-browser-dynamic';
import {AppModule} from './app.module';

setTimeout(
  () => platformBrowserDynamic().bootstrapModule(AppModule),
  2000
)
```

When you start the app, the nested elements within **<acme>** are displayed, because the browser handles this non-standard HTML tag as if it were a **<div>** (Figure 7-8).

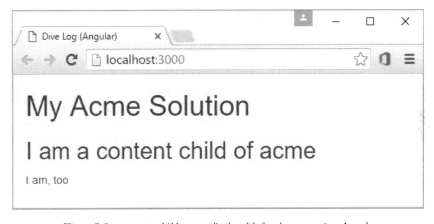

Figure 7-8: <acme> children are displayed before bootstrapping Angular

However, as soon as the application has been bootstrapped, the framework removes the content children of **<acme>**, and renders **AcmeComponent** according to its template (Figure 7-9). The DOM in the browser confirms this fact, too (Figure 7-10).

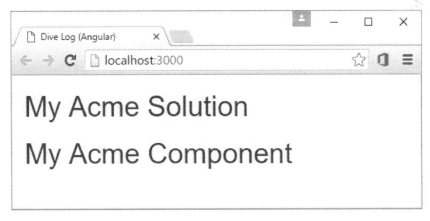

Figure 7-9: The content children of <acme> are removed

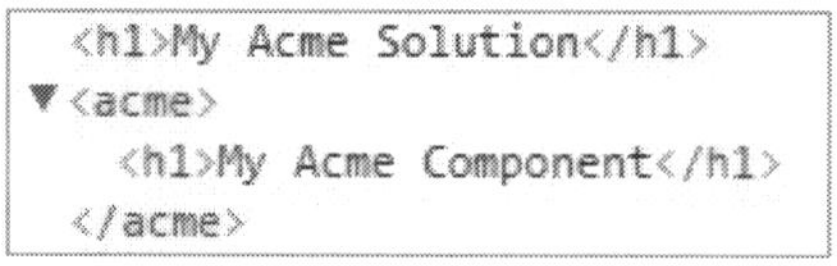

```
<h1>My Acme Solution</h1>
▼ <acme>
    <h1>My Acme Component</h1>
  </acme>
```

Figure 7-10: The DOM fragment of <acme>

Accessing Content Children with <ng-content>

We intend to transpose the content children into the view so that the content can be displayed. We can do it three ways:

#1. We add an **<ng-content>** element to the template of your component, and Angular will replace **<ng-content>** with the content children.

#2. We can apply **<ng-content>** with the **select** attribute that declares a CSS selector to transpose a subset of the content children to the view.

#3. We can access the content children within the component and define the logic to render the view.

The first two methods are straightforward; the third one provides flexibility.

The **Exercise-07-09** folder demonstrates the first option. This sample bootstraps **SuperbComponent** (Listing 7-14).

Listing 7-14: superb.component.ts (Exercise-07-09)

```typescript
import {Component} from '@angular/core';

@Component({
  selector: 'superb',
  template: `
    <h1>My SuperbComponent</h1>
    <acme>
      <h3>I am a content child of acme</h3>
      <p>I am, too</p>
    </acme>
  `
})
export class SuperbComponent {
}
```

Here, **<acme>** has content children. As Listing 7-15 shows, **AcmeComponent** adds **<ng-content>** to its template, and so the **<h3>** and **<p>** content children are transposed to the view (Figure 7-11).

Listing 7-15: acme.component.ts (Exercise-07-09)

```
import {Component} from '@angular/core';

@Component({
  selector: 'acme',
  template: `
    <h2>My Acme Component</h2>
    <ng-content></ng-content>
  `
})
export class AcmeComponent {
}
```

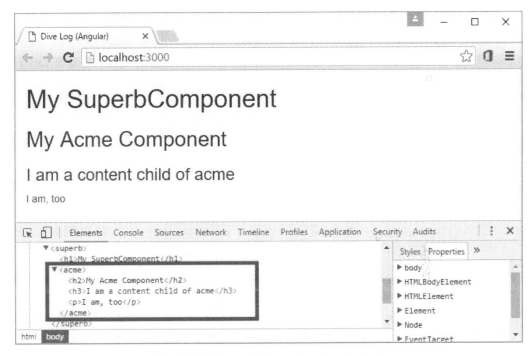

Figure 7-11: Content children are transposed to the view

The **select** property of **ng-content** adds more flexibility, as the sample in the **Exercise-07-10** folder demonstrates. This app demonstrates using a few component lifecycle hooks, too, namely **OnInit**, **AfterContentInit**, and **AfterViewInit**. Each of these events logs messages through a component, **MessageBoardComponent**, which is added with the **<yw-messages>** tag to the markup.

Listing 7-16 shows that **<yw-messages>** has a content with a **<header>** and a **<footer>** tag.

Listing 7-16: app.component.ts (Exercise-07-10)

```
import {Component} from '@angular/core';
import {OnInit, AfterContentInit, AfterViewInit} from '@angular/core';

@Component({
  selector: 'yw-app',
  template: `
    <div class="container-fluid">
      <yw-messages [messages]="logMessages">
        <header>
          <h2>Messages Logged</h2>
        </header>
        <footer>
          <p>--- End of messages</p>
        </footer>
      </yw-messages>
    </div>
  `
})
export class AppComponent implements OnInit,
  AfterContentInit, AfterViewInit {

    logMessages: string[] = [];
    count = 0;

    ngOnInit() {
      this.log('ngOnInit');
    }

    ngAfterContentInit() {
      this.log('ngAfterContentInit');
    }

    ngAfterViewInit() {
      this.log('ngAfterViewInit');
    }

    log(message: string) {
      this.logMessages.push(`${++this.count}: ${message}`);
    }
}
```

Take a look at Listing 7-17. Here, the `<ng-content>` tag applies the **select** attribute to transpose `<header>` and `<footer>` from the content of `<yw-messages>` to its view.

Listing 7-17: message-board.component.ts (Exercise-07-10)

```
import {Component} from '@angular/core';

@Component({
  selector: 'yw-messages',
  template: `
    <div>
      <ng-content select="header"></ng-content>
      <ul>
        <li *ngFor="let message of messages">
          {{message}}
        </li>
      </ul>
      <ng-content select="footer"></ng-content>
    </div>
  `,
  inputs: ['messages']
})
export class MessageBoardComponent {
  messages: string[] = [];
}
```

As this listing demonstrates, you can add multiple `<ng-content>` tags to a component template.

When you run the app, it displays the messages wrapped with the specified header and footer (Figure 7-12).

Figure 7-12: The MessageBoard app in action

Besides presenting the **select** attribute, this sample also demonstrates the order of the three lifecycle events we utilized in the code:

Angular invokes the **OnInit** hook when it has created and initialized the component and checked all of the data-bound properties the first time. When the framework calls **OnInit**'s hook method, **ngOnInit()**, it has not checked the children of the component yet.

When the content is fully initialized, Angular calls the **AfterContentInit** hook (its method is **ngAfterContentInit**).

When the component's view is fully initialized—the template is prepared, the content is transposed—, the framework invokes **ngAfterViewInit**, which is the hook method of **AfterViewInit**.

NOTE: Other lifecycle events that happen while the component's view is displayed. Soon, you will learn about them.

Accessing Content from the Component Code

We often need to transform the content dynamically, and that would be almost impossible or very tedious with **<ng-content>**. Angular provides access to a component's content with the help of the **@ContentChild()** and **@ContentChildren()** annotations.

Assume, we want to create a **TocComponent**, which can display a table of contents for a book with a dynamically set **level** property, which specifies the number of levels we intend to show. The **Exercise-07-11** folder contains a sample app that implements such a component. Listing 7-18 shows the markup that sets up **TocComponent**.

Listing 7-18: app.template.html (Exercise-07-11)

```
...
<yw-toc [level]="currentLevel">
  <header>Table of Contents</header>
  <h1>Introduction</h1>
  <h2>Scuba Basics</h2>
  <h3>Volume</h3>
  <h3>Pressure</h3>
  <h3>Depth</h3>
  <h2>The Environment</h2>
  <h2>Open Water Certifications</h2>
  <h3>Open Water Diver</h3>
  <h3>Advanced Open Water Diver</h3>
  <h3>Rescue Diver</h3>
  <h3>Dive Master</h3>
</yw-toc>
...
```

The running application transforms the content of **<yw-toc>** so that it displays only the specified levels. For example, when we select the first two levels, **<h3>** elements are omitted from the view (Figure 7-13).

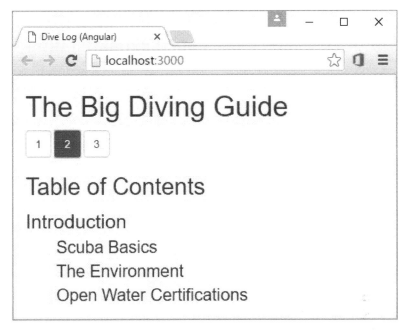

Figure 7-13: The TocComponent sample in action

As shown in Listing 7-19, `TocComponent` relies on `@ContentChild()` and `@ContentChildren()` to access the content from code.

Listing 7-19: app.template.html (Exercise-07-11)

```
import {Component, ElementRef} from '@angular/core';
import {ContentChild, ContentChildren, QueryList} from '@angular/core';
import {HnDirective} from './hn.directive';
import {HeaderDirective} from './header.directive';

@Component({
  selector: 'yw-toc',
  template: `
    <div>
      <h2>{{header.text}}</h2>
      <div *ngFor="let element of getFilteredElements()"
        [style.padding-left.px]="40*(element.level-1)"
        [style.font-size.em]="1.8-(element.level-1)*0.25">
        {{element.text}}
      </div>
    </div>
  `,
  inputs: ['level']
})
export class TocComponent {
  @ContentChild(HeaderDirective) header;
  @ContentChildren(HnDirective) elements: QueryList<HnDirective>;
```

```
  level = 2;

  getFilteredElements() {
    return this.elements.toArray().filter(e => e.level <= this.level);
  }
}
```

These decorations accept either a string or a directive/component type. **@ContentChildren()** instructs the framework to put the list of *all* content children that match with the annotation's argument into the property it decorates. These children are represented with a **QueryList<>** generic instance. **QueryList<>** is an immutable, iterable list and Angular keeps it up-to-date. Moreover, we can subscribe to the changes of the list.

@ContentChild() obtains the *first* content child that matches with its argument.

> *NOTE: When we use a string with these decorations, Angular matches them with a local template variable, and retrieves the first child/ all children of the element the matching local variable represents.*

In this sample, we instruct the **@ContentChild(HeaderDirective)** definition to retrieve that child of **TocComponent**, which represents the **<header>** tag. Listing 7-20 shows how simple the declaration of **HeaderDirective** is.

Listing 7-20: header.directive.ts (Exercise-07-11)

```
import {Directive, ElementRef, OnInit} from '@angular/core';

@Directive({
  selector: 'header'
})
export class HeaderDirective implements OnInit {
  text: string;
  constructor(private element: ElementRef) {
  }

  ngOnInit() {
    this.text = this.element.nativeElement.textContent;
  }
}
```

As we expect, this directive matches with the **<header>** tag through its selector. When Angular invokes the constructor—as you already learned—it injects the host element wrapper (**ElementRef**) of the directive, which happens to be **<header>** itself. We need to wait for the **OnInit** lifecycle event to be sure that the engine has prepared the child element. When that is initialized, we store its **textContent** into **text**.

We use the same pattern—with a little tweak—to declare **HnDirective** to access all heading level tags (Listing 7-21).

Listing 7-21: hn.directive.ts (Exercise-07-11)

```
import {Directive, ElementRef, OnInit} from '@angular/core';

@Directive({
  selector: 'h1, h2, h3, h4, h5, h6'
})
export class HnDirective implements OnInit {
  level: number;
  text: string;
  constructor(private element: ElementRef) {
  }

  ngOnInit() {
    this.level =
parseInt(this.element.nativeElement.nodeName.substring(1));
    this.text = this.element.nativeElement.textContent;
  }
}
```

The highlighted selector makes this directive match with all HTML heading tags. In the **ngOnInit()** method, we extract the **level** and **text** information from the heading.

With the help of **HeaderDirective** and **HnDirective** we can use the content of **TocComponent** to render its view:

```
@Component({
  selector: 'yw-toc',
  template: `
    <div>
      <h2>{{header.text}}</h2>
      <div *ngFor="let element of getFilteredElements()"
        [style.padding-left.px]="40*(element.level-1)"
        [style.font-size.em]="1.8-(element.level-1)*0.25">
        {{element.text}}
      </div>
    </div>
  `,
  inputs: ['level']
})
export class TocComponent {
  @ContentChild(HeaderDirective) header;
  @ContentChildren(HnDirective) elements: QueryList<HnDirective>;
  level = 2;

  getFilteredElements() {
    return this.elements.toArray().filter(e => e.level <= this.level);
  }
}
```

A Few Remarks On Directives

`HeaderDirective` and `HnDirective` are special attribute directives. They are attached to native HTML tags, but they do not modify the behavior of them. Their mere purpose is to extract information from the corresponding DOM elements and map them to properties.

Because these directives are included in the root application module, they add a little extra effect. Although we intend them to map only the HTML heading tags within `<yw-toc>`, they are attached to every heading element independently of where they are.

The `Exercise-07-12` folder contains a sample that logs whenever any of the directives attaches itself to an HTML element. This app catches the mentioned effect, as Figure 7-14 indicates. The arrows sign the extra heading elements outside of `<yw-toc>` to which the framework attaches `HnDirective`.

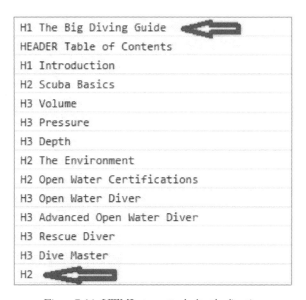

Figure 7-14: HTML tags attached to the directives

Because our directives do not add extra behavior to the HTML elements, this side effect does not have observable issues. Of course, it adds a very slight performance overhead, but that is insignificant.

Nonetheless, in real applications, such a mapping of native HTML elements might cause unwanted behavior or performance issues, so you must be circumspect.

Mapping Component Content

You can nest components into other components' content. For example, you can define a menu with sections and items, as shown in Listing 7-22.

Listing 7-22: app.template.html (Exercise-07-13)

```html
<div class="container-fluid">
  <h1>Scuba Diving News</h1>
  <div [style.width.px]="200">
    <yw-menu #menu>
      <menu-section title="Red Sea">
        <menu-item title="News #1"></menu-item>
        <menu-item title="News #2"></menu-item>
        <menu-item title="News #3"></menu-item>
      </menu-section>
      <menu-section title="Caribbean">
        <menu-item title="News #4"></menu-item>
        <menu-item title="News #5"></menu-item>
        <menu-item title="News #6"></menu-item>
      </menu-section>
      <menu-section title="Pacific">
        <menu-item title="News #7"></menu-item>
        <menu-item title="News #8"></menu-item>
        <menu-item title="News #9"></menu-item>
      </menu-section>
    </yw-menu>
  </div>
  <h3>Selected: {{menu.selectedTitle}}</h3>
</div>
```

The completed menu displays only a single—active—section with the corresponding items. When we click an item, we can query its title with the menu's **selectedTitle** property (Figure 7-15).

Obviously, we need to create a component for the **<yw-menu>** tag. What if we implemented the objects behind **<menu-section>** and **<menu-item>** with components, too? In this case, we could use these components for two purposes: to represent the content of the menu and implement the logic that would build up the menu's view.

Building a Blueprint

Let's implement this app. First, we create a skeleton of the menu with its sections and items. The blueprints of the three components are shown in Listing 7-23, Listing 7-24, and Listing 7-25.

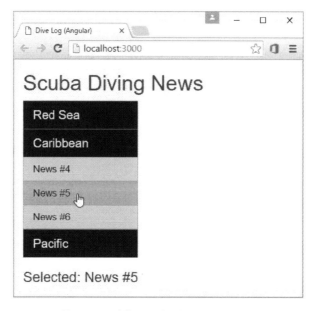

Figure 7-15: The completed menu component

Listing 7-23: menu-item.component.ts (Exercise-07-13)

```
...
@Component({
  selector: 'menu-item',
  template: `
    <div class="title">
      {{title}}
    </div>
  `,
  inputs: ['title'],
})
export class MenuItemComponent {
  title: string;
}
```

Listing 7-24: menu-section.component.ts (Exercise-07-13)

```
...
@Component({
  selector: 'menu-section',
  template: `
    <div class="title">
      <h3>{{title}}</h3>
      <menu-item *ngFor="let item of items"
        [title]="item.title">
      </menu-item>
    </div>
  `,
```

```
    inputs: ['title']
})
export class MenuSectionComponent {
  title: string;
  @ContentChildren(MenuItemComponent) items: QueryList<MenuItemComponent>;
}
```

Listing 7-25: menu.component.ts (Exercise-07-13)

```
...
@Component({
  selector: 'yw-menu',
  template: `
    <div>
      <menu-section *ngFor="let section of sections"
        [title]="section.title">
      </menu-section>
    </div>
  `
})
export class MenuComponent {
  selectedTitle: string;

  @ContentChildren(MenuSectionComponent) sections:
QueryList<MenuSectionComponent>;
}
```

As you see, the code uses the **@ContentChildren()** annotation to collect the items of a section (Listing 7-24) and sections of a menu (Listing 7-25). According to what you learned, this code should be just enough to display the menu with its sections and items.

When you run the blueprint app, something unexpected happens. Only the menu sections are visible. Items do not show up (Figure 7-16)—and there is no error message in the console output.

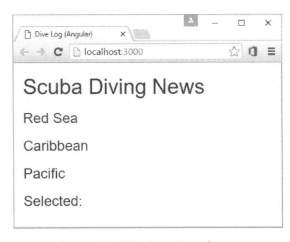

Figure 7-16: Menu items do not show up

To understand why we see what we see, Figure 7-17 gives a clue. For the sake of simplicity, this figure depicts only one section with two items.

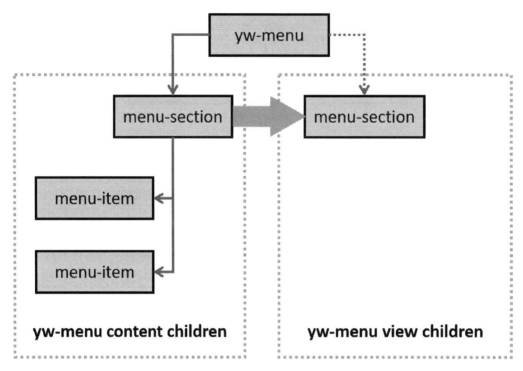

Figure 7-17: Transposing the menu-section content child

On the left, you can see the content children Angular creates for **<yw-menu>**. Due to the **@ContentChildren()** annotation on **MenuComponent** and **MenuSectionComponent**, the framework creates one **MenuSectionComponent** and two **MenuItemComponent** instances. However, these are not displayed, for they only represent the content children.

The big gray arrow in the middle of the figure represents the transposal the template of **MenuComponent** carries out:

```
<menu-section *ngFor="let section of sections"
  [title]="section.title">
</menu-section>
```

Here, the **sections** property represents the list of **<menu-section>** content children. This markup creates a new **MenuSectionComponent** instance for each content child and copies the child's title into the new component. Thus the two **menu-section** boxes in the figure represent two separate component instances: the one the left is a content child, the other on the right is a view child. Because in the markup above we do not create any **MenuItemComponent** instances, no one is added to the view.

Fixing the Templates

To resolve the issue, we need to add extra markup lines to the **MenuItemComponent** and **MenuSectionComponent** templates:

```
#1: MenuComponent template:

<div>
  <menu-section *ngFor="let section of sections"
    [title]="section.title">
    <menu-item *ngFor="let item of section.items"
      [title]="item.title">
    </menu-item>
  </menu-section>
</div>

#2: MenuSectionComponent template:

<div class="title">
  <h3>{{title}}</h3>
</div>
<ng-content></ng-content>
```

Modification #1 clones the **MenuItemComponent** children from the **<yw-toc>** content into the **MenuComponent** template. These children become content children of **<menu-section>** in the **MenuComponent** template. Modification #2 takes care that all these **menu-item** content children become view children, and thus they are displayed. Figure 7-18 depicts the result of these transformations. The boxes with dark shades show the view children that finally get rendered within the view of **MenuComponent**.

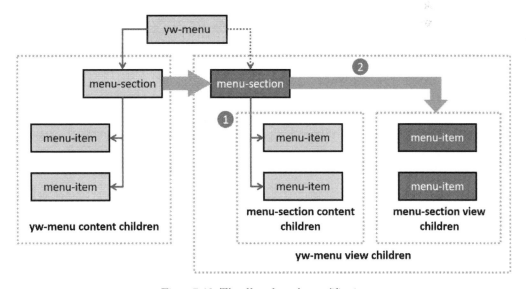

Figure 7-18: The effect of template modifications

You can find the sample with these fixes in the **Exercise-07-14** folder. When you run the modified app, sections and items are both displayed. You can find the completed app with all CSS stylings and event handling in the **Exercise-07-15** folder.

Using Directives for Content Children

We could have avoided the whole mess coming from the dual role of **MenuSectionComponent** and **MenuItemComponent**, if we had used directives to represent **menu-section** and **menu-item**. To show you that this approach is more straightforward, I implemented **MenuComponent** that way (**Exercise-07-16**). Without any further explanations, take a look at Listing 7-26 and Listing 7-27, which show the directive classes.

Listing 7-26: menu-section.directive.ts (Exercise-07-16)

```
import {Component} from '@angular/core';
import {ContentChildren, QueryList} from '@angular/core';

import {MenuItemComponent} from './menu-item.component';

@Component({
  selector: 'menu-section',
  template: `
    <div class="title">
      <h3>{{title}}</h3>
      <menu-item *ngFor="let item of items"
        [title]="item.title">
      </menu-item>
    </div>
  `,
  inputs: ['title']
})
export class MenuSectionComponent {
  title: string;
  @ContentChildren(MenuItemComponent) items: QueryList<MenuItemComponent>;
}
```

Listing 7-27: menu-item.directive.ts (Exercise-07-16)

```
import {Directive} from '@angular/core';
import {ContentChildren, QueryList} from '@angular/core';

@Directive({
  selector: 'menu-item',
  inputs: ['title'],
})
export class MenuItemDirective {
  title: string;
}
```

With these directives, the template of **MenuComponent** is short and self-explaining (Listing 7-28).

Listing 7-28: menu.template.html (Exercise-07-16)

```
<div>
  <menu-section *ngFor="let section of sections"
    [title]="section.title">
    <menu-item *ngFor="let item of section.items"
      [title]="item.title">
    </menu-item>
  </menu-section>
</div>
```

Keep in mind that using the same component type to represent content and view at the same time has some challenges. Always think it over how you would describe the content and create the view. Often it is worth to create directives to represent the content and make separate components that process the content directives.

Working with View Children

Just as we can query content children in the code, we can access view children with the help of the **@ViewChild()** and **@ViewChildren()** decorators. These decorators can be used similarly as **@ContentChild()** and **@ContentChildren()**, but of course, they work on the component view and not on the content.

> **NOTE**: *Just for a short recap: view children are the children element located within the component template.*

When our component renders a collection of children component of the same type, we rarely need to use **@ViewChildren()** to access them—we have already a property that allows accessing those children directly.

A more practical scenario is when we need to get a reference to particular view child component instance so that we can access its properties and methods from the hosting component's code.

Assume, we update the simple dive log application so that we can start quick queries with buttons. When we click a button, it works as if we typed some text in the search box. For example, as Figure 7-19 shows, when we click the Search for '8' button, the app works as if we typed '8' in the search box.

You can find this application in the **Exercise-07-17** folder. It uses the same code you already met several times in the book, but this sample implements a separate component that represents the search box at the top right part of the figure (Listing 7-29).

Figure 7-19: The modified dive log app

Listing 7-29: search-box.component.ts (Exercise-07-17)

```
import {Component, EventEmitter} from '@angular/core';

@Component({
  selector: 'search-box',
  template: `
    <input #input class="form-control input-lg"
      placeholder="Search"
      [value]=searchText
      (keyup)="searchText=input.value" />
  `,
  inputs: ['searchText']
})
export class SearchBoxComponent {
  searchText: string;
}
```

SearcBoxComponent exposes a property, **searchText** that can be used not only to get the value typed into the box but also for setting it. The **DiveLogComponent** template applies **SearchBoxComponent** and uses a template reference variable to implement the quick search functionality (Listing 7-30).

Listing 7-30: dive-log-template.html (Exercise-07-17)

```
<div class="row">
  <div class="col-sm-5">
    <button class="btn btn-default"
      (click)="searchBox.searchText='8'">
        Search for '8'
    </button>
```

```
    <button class="btn btn-default"
      (click)="searchBox.searchText='Hun'">
        Search for 'Hun'
    </button>
    <button class="btn btn-default"
      (click)="searchBox.searchText=''">
        Clear
    </button>
  </div>
  <div class="col-sm-4 col-sm-offset-3">
    <search-box #searchBox></search-box>
  </div>
</div>
<div class="row">
  <div class="col-sm-4"
    *ngFor="let dive of dives | contentFilter:searchBox.searchText">
    <h3>{{dive.site}}</h3>
    <h4>{{dive.location}}</h4>
    <h2>{{dive.depth}} feet | {{dive.time}} min</h2>
  </div>
</div>
```

This approach with the template reference variable works unless we need to access the
SearchComponent instance from within **DiveLogComponent** itself. For example, if we need to log
every quick search or the text for a particular search is assembled in the component, we need to
find another way.

@ViewChildren() will help us in this scenario. Listing 7-31 shows how we can leverage it in
DiveLogComponent.

Listing 7-31: dive-log.component.ts (Exercise-07-18)

```
import {Component, AfterViewInit} from '@angular/core';
import {ViewChild} from '@angular/core';
import {DiveLogEntry} from './dive-log-entry';
import {SearchBoxComponent} from './search-box.component';

@Component({
  selector: 'divelog',
  templateUrl: 'app/dive-log.template.html'
})
export class DiveLogComponent implements AfterViewInit {
  @ViewChild(SearchBoxComponent) searchBox: SearchBoxComponent;
  dives = DiveLogEntry.StockDives;

  searchFor(key: string) {
    this.searchBox.searchText = key;
    console.log(`New search text: ${this.searchBox.searchText}`)
  }
```

```
ngAfterViewInit() {
    if (this.searchBox) console.log('searchBox initialized.')
}
}
```

The framework takes care of assigning the first **SearchBoxComponent** view child to the **searchBox** property. Angular invokes the **AfterViewInit** lifecycle hook when the view is initialized, and at that moment we can access **searchBox**. If we tried to use it in an earlier stage of the component's life, the property value would be undefined.

DivLogComponent provides the **searchFor()** method to set the **searchText** property of the child **SearchBoxComponent**. As Listing 7-32 shows, we do not need a template reference variable anymore.

Listing 7-32: dive-log-template.html (Exercise-07-18)

```
<div class="row">
  <div class="col-sm-5">
    <button class="btn btn-default"
      (click)="searchFor('8')">
        Search for '8'
    </button>
    <button class="btn btn-default"
      (click)="searchFor('Hun')">
        Search for 'Hun'
    </button>
    <button class="btn btn-default"
      (click)="searchFor('')">
        Clear
    </button>
  </div>
  <div class="col-sm-4 col-sm-offset-3">
    <search-box></search-box>
  </div>
</div>
<div class="row">
  <div class="col-sm-4"
    *ngFor="let dive of dives | contentFilter: searchBox.searchText">
    <h3>{{dive.site}}</h3>
    <h4>{{dive.location}}</h4>
    <h2>{{dive.depth}} feet | {{dive.time}} min</h2>
  </div>
</div>
<hr/>
```

The **contentFilter** pipe still uses the **searchBox.searchText** expression, but this time **searchBox** is not a template reference variable: it is a property of **DiveLogComponent**.

Lifecycle Hooks

You have already learned about a couple of lifecycle hooks. Moreover, you applied a few in the examples you studied earlier. Now, it is time to get acquainted with all of them.

While managing the lifecycle of directives and components, Angular provides hook methods so that we can act when these particular events happen. We can implement a hook method if we intend to respond to the corresponding lifecycle event.

When it is the time to involve a hook, the framework just checks if the directive or component defines the hook method. Provided it exists, Angular invokes it.

For example, if we want to obtain information about input property changes in our `AcmeComponent`, we add the `ngOnChanges()` method to it:

```
@Component({
  selector: 'acme',
  // ...
})
export class AcmeComponent {
  // ...
  ngOnChanges() {
    // ...
  }
// ...
}
```

Angular defines TypeScript interfaces for each lifecycle hook. These interfaces express the intention of a component developer—and besides allows the compiler more thorough checks—, so I suggest you implement them in the affected component and directive classes.

The modified implementation of `AcmeComponent` helps readability and thus code maintenance because the declaration of the class tells that it manages the `OnChanges` lifecycle hook:

```
...
import {OnChanges } from '@angular/core';

@Component({
  selector: 'acme',
  // ...
})
export class AcmeComponent implements OnChanges {
  // ...
  ngOnChanges() {
    // ...
  }
// ...
}
```

Every lifecycle hook interface has exactly one method that has the name of the hook with an "**ng**" prefix.

A Lifecycle Hooks Sample

In the **Exercise-07-19** folder, you can find a modified version of the **MenuComponent** you created earlier in this chapter. This sample implements all Angular lifecycle hooks in the component class. The UI allows creating and destroying a menu instance, and provides an extra button, Toggle color, to demonstrate input property changes (Figure 7-20).

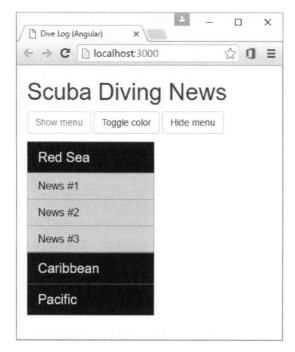

Figure 7-20: The modified menu sample

Let's dive into the details of lifecycle hooks!

The app adds a message to the console output for each hook method invoked. These messages display a sequence number and the name of the hook. We use the output to check the order of them.

When we start the app, no menu is displayed. Clicking the Show menu button logs these seven messages:

```
1: OnChanges
2: OnInit
3: DoCheck
4. AfterContentInit
5: AfterContentChecked
```

```
6: AfterViewInit
7: AfterViewChecked
```

Due to this **ngIf** directive, no **MenuComponent** exists unless we click the Show menu button:

```
<yw-menu #menu *ngIf="displayMenu"
  [useDefaultColor]="defaultColor">
  <!-- ... -->
</yw-menu>
```

OnChanges is the first hook invoked. The framework calls this hook whenever it detects a change to any input properties. After creating a new component instance (Angular uses the **new** operator), the framework sets the **useDefaultColor** input property of **MenuComponent**, and this action triggers **OnChanges**.

> *NOTE: As you already learned, in a single change detection cycle Angular may detect multiple input property changes. All of them is passed as an argument of **ngOnChanges()**.*

The second hook is **OnInit**. Angular invokes this hook when all data-bound properties are initialized. In this example, we have only one input property, **useDefaultColor**.

In *Chapter 5, The Bootstrap Process*, you learned that Angular invokes a change detection cycle as soon as the component has been bootstrapped. Every change detection cycle invokes the **DoCheck** hook so that you can add supplement code to change detection—implement those checks that cannot be observed by Angular.

> *NOTE: You need to implement **ngDoCheck()** only very rarely, in unusual scenarios. Practically, Angular detects every change.*

MenuComponent has content children. The framework invokes **AfterContentInit** when it has successfully projected the component's content into the view. This event is directly followed by **AfterContentChecked** to sign that Angular has checked all the bindings of content it projected into the view.

After the content's projection, Angular creates components for the view children. When this step has completed, the framework invokes the **AfterViewInit** hook. As soon as the bindings in these child components are checked, **AfterViewChecked** is called.

Now, when we click the Toggle color button, the following four new message is appended to the log:

```
8: OnChanges
9: DoCheck
10: AfterContentChecked
11: AfterViewChecked
```

The click changes the **useDefaultColor** input property of **MenuComponent**—and this change triggers **OnChange**, and then **DoCheck**. The highlighted part of the component's template utilizes **useDefaultColor**:

```
<div class="sectionTitle"
  [class.sectionLast]="1"
  [class.maroonSection]="useDefaultColor"
  (click)="selectSection(section)">
  {{section.title}}
</div>
```

Because of the value of the property changes, Angular updates data binding. This update triggers the **AfterContentChecked**, and then **AfterViewChecked**.

When we click the Hide menu button, the condition of **ngIf** becomes false:

```
<yw-menu #menu *ngIf="displayMenu"
  [useDefaultColor]="defaultColor">
  <!-- ... -->
</yw-menu>
```

According to its definition, **ngIf** removes the **<yw-menu>** tag from the DOM. We do not need the **MenuComponent** in the UI—not unless you click Show menu again—so Angular invokes the **OnDestroy** hook to clean up the resources held by the component:

```
12: OnDestroy
```

When the hook method returns, the framework destroys the component.

A Few Remarks on Lifecycle Hooks

All components are directives. When you create custom directives, you can implement the **OnInit**, **OnChanges**, **DoCheck**, and **OnDestroy** hooks.

The only hook method that accepts parameters is **ngOnChanges**; the others are parameterless.

Other Angular extension modules, such as the Component Router or Angular Forms, or future extensions may add their extra lifecycle hooks. We can also create our component libraries with custom lifecycle hooks.

Summary

A component represents a part of the UI and controls its behavior. The logic behind the component determines the data displayed in the corresponding UI and responds to user actions. Components never manipulate the DOM directly; they use templates with binding to declare and update the UI indirectly.

We do not need to create Angular components. The framework does it for us, and it manages their entire lifecycle. With defining hook methods, we can respond to the events of the lifecycle.

Components can form parent-child hierarchy, where the nodes can communicate with other nodes. With input properties, a parent can pass data down to its children. A child can use output event binding to notify its parent.

Unrelated components can use services as intermediaries to establish unidirectional or bidirectional communication.

We can utilize content children—elements between the opening and closing tags of our component's host element—to create the view of the component.

Chapter 8: Form Management

Creating template-driven forms
Using two-way binding with `ngModel`
Adding validation to forms
Providing visual validation feedback
Understanding Reactive Forms
Adding custom validation to forms

We cannot avoid using we forms. We log in to websites, order from a web shop, prepare and send an email, add comments to posts. As developers of web apps, we have to put together HTML forms so that users can enter and submit data for further processing.

Designing a form that provides excellent user experience is not easy—not at all. Besides the traditional UX designer skills, we have to deal with many subtle things such as populating the form with data, validation, change tracking, data submission, error handling and so on. Angular does not add too much to your designer skills, but the framework helps you implement the nitty-gritty details that form management requires.

In this chapter, you will learn the fundamental concepts of Angular form management. Instead of forcing you to use a single way of handling forms, Angular allows you to choose from two methods:

#1. You can create forms with the template-driven approach. You create a template and assign validation attributes and directives to it to declare your intentions to control the form. When you run the app, Angular interprets the template and derives its form control model.

#2. You can define your forms with the Reactive Forms approach. You still create a template for the form, but you set up the form control model in code.

You will learn about both methods through examples.

Template-Driven Forms

When we create a form, we have common tasks to carry out so that we can process the collected data. Users appreciate the guidance that helps them understand what we expect from them. As they fill in a form, we need to validate the data and give them feedback when they specify something we cannot understand or accept.

Building a form contains many repetitive tasks, such as populating the form with data, checking if a change completes with valid information, altering the visual properties of the form to warn the user, managing the data submission, and so on.

With the help of template-driven forms, Angular handles the majority of these tasks.

The Dive Log Entry Form

To demonstrate these concepts, we are going to create a form that allows a scuba diver to enter a dive log record. Figure 8-1 shows this app in action. The logic behind the form checks for validation issues and displays them.

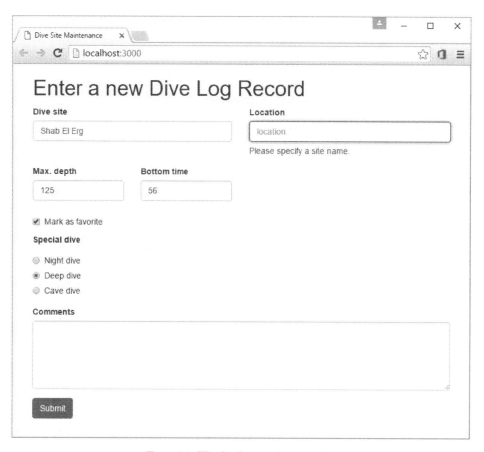

Figure 8-1: The dive log entry form in action

When the user submits the form, the app displays the submitted content (Figure 8-2).

The longest part of the app we are going to build is the HTML template that displays the form (Listing 8-1).

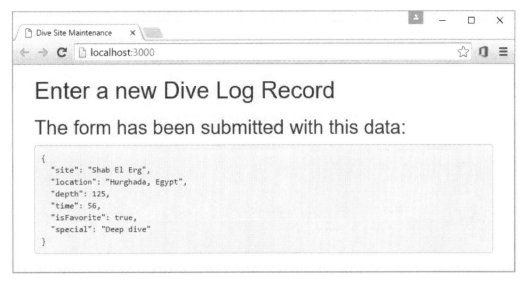

Figure 8-2: The submitted form

Listing 8-1: dive-log-form.template.html (Exercise-08-01)

```html
<form>
  <div class="row">
    <div class="col-sm-6">
      <div class="form-group">
        <label for="site">Dive site</label>
        <input class="form-control" id="site"
          placeholder="dive site">
      </div>
    </div>
    <div class="col-sm-6">
      <div class="form-group">
        <label for="location">Location</label>
        <input class="form-control" id="location"
          placeholder="location">
      </div>
    </div>
  </div>
  <div class="row">
    <div class="col-sm-3">
      <div class="form-group">
        <label for="depth">Max. depth</label>
        <input class="form-control" id="depth">
      </div>
    </div>
```

```
        <div class="col-sm-3">
          <div class="form-group">
            <label for="time">Bottom time</label>
            <input class="form-control" id="time">
          </div>
        </div>
      </div>
      <div class="checkbox">
        <label>
          <input type="checkbox" name="favorite" checked>
          Mark as favorite
        </label>
      </div>
      <div class="form-group">
        <label>Special dive</label>
        <div class="radio"
          *ngFor="let diveType of specialDives">
          <label>
            <input type="radio" name="special" [value]="diveType">
            {{diveType}}
          </label>
        </div>
      </div>
      <div class="form-group">
        <label for="comments">Comments</label>
        <textarea class="form-control" id="comments" rows="5">
        </textarea>
      </div>
      <button class="btn btn-primary" type="submit">
        Submit
      </button>
    </form>
```

In the markup, I highlighted those parts that represent the controls of the form. The remaining code provides the layout. Just as in the previous examples, this app uses Bootstrap. The **<div>** wrapper sections ensure that the controls are rendered with their corresponding labels (**form-group** and **form-control** classes), and a few of them shares the same row (**row**, **col-sm-6**, and **col-sm-3** classes).

The component behind the form has only a **specialDives** property to define the options for the radio buttons of the form (Listing 8-2).

Listing 8-2: dive-log-form.component.ts (Exercise-08-01)

```
import {Component} from '@angular/core';

@Component({
  selector: 'dive-log-form',
  templateUrl: 'app/dive-log-form.template.html'
})
```

```
export class DiveLogFormComponent {
  specialDives = [
    "Night dive",
    "deep dive",
    "Cave dive"
  ]
}
```

The application root module is as simple as we expect (Listing 8-3).

Listing 8-3: app.module.ts (Exercise-08-01)

```
import {NgModule} from '@angular/core';
import {BrowserModule} from '@angular/platform-browser';

import {AppComponent} from './app.component';
import {DiveLogFormComponent} from './dive-log-form.component'

@NgModule({
  imports: [BrowserModule],
  declarations: [
    AppComponent,
    DiveLogFormComponent
  ],
  bootstrap: [AppComponent]
})
export class AppModule { }
```

Establishing Two-Way Binding

In the previous chapters, we used the `<input>` control several times. For example, in Chapter 3, we applied this markup to process the content of the site name information:

```
<input #siteNameBox class="form-control input-lg" type="text"
  [value]="siteName"
  placeholder="site name"
  (keyup)="siteName=siteNameBox.value" />
```

This markup—with the help of the host component—ensures that the site name information entered by the user is bound to the **siteName** property of the component class as the user types in the text. Due to the **[value]="siteName"** assignment, the initial content of the input box is set to **siteName**. To access the current value in the **keyup** event, we have to define the **siteNameBox** template reference variable.

Should this be the mandatory dance to declare each single form control, we would have to type a lot.

283

In *Chapter 5, Two-Way Binding*, you learned about a special notation, **[(ngModel)]**:

```
<input class="form-control input-lg" type="text"
  [(ngModel)]="siteName" />
```

This syntax defines *two-way binding*. When the user types into the **<input>** control, the text gets into **siteName**. Vice versa, if the component logic changes **siteName**, the new value appears in the text box.

Let's update the dive log entry app to utilize two-way bindings!

The core of two-way binding is the **ngModel** directive, which is located in the **FormsModule** NgModule. To apply **ngModel**, we need to import **FormsModule**, as shown in Listing 8-4.

Listing 8-4: app.module.ts (Exercise-08-02)

```
import {NgModule} from '@angular/core';
import {BrowserModule} from '@angular/platform-browser';
import {FormsModule} from '@angular/forms';

import {AppComponent} from './app.component';
import {DiveLogFormComponent} from './dive-log-form.component'

@NgModule({
  imports: [
    BrowserModule,
    FormsModule
  ],
  declarations: [
    AppComponent,
    DiveLogFormComponent
  ],
  bootstrap: [AppComponent]
})
export class AppModule { }
```

We could add properties to the **DiveLogFormComponent** to hold form data:

```
export class DiveLogFormComponent {
  site: string;
  location: string;
  depth: number;
  time: number;
  isFavorite: boolean;
  special: string;
  // ...
}
```

Nonetheless, we want to handle these properties together, as they represent a model of the form. We will use the form data in other parts of the app, too. Thus, we better enclose them into their dedicated type, `DiveLogEntry`, as shown in Listing 8-5.

Listing 8-5: dive-log-entry.ts (Exercise-08-02)

```
export interface DiveLogEntry {
  site: string;
  location: string;
  depth: number;
  time: number;
  isFavorite: boolean;
  special: string;
  comments?: string;
}
```

As you see, here, `DiveLogEntry` is defined as an interface—though I might have declared it as a class. A TypeScript interface defines the *shape* of an object—just the definition—, while a class is also an implementation. You do not make any mistake if you use a class. Here, I found an interface is more flexible—with the future increments of the app in my mind.

We can add an entry property to `DiveLogFormComponent` to represent the current form data (Listing 8-6).

Listing 8-6: dive-log-form.component.ts (Exercise-08-02)

```
import {Component} from '@angular/core';
import {DiveLogEntry} from './dive-log-entry';

@Component({
  selector: 'dive-log-form',
  templateUrl: 'app/dive-log-form.template.html'
})
export class DiveLogFormComponent {
  specialDives = [
    "Night dive",
    "Deep dive",
    "Cave dive"
  ]

  entry: DiveLogEntry = {
    site: 'Shab El Erg',
    location: 'Hurghada, Egypt',
    depth: 125,
    time: 56,
    isFavorite: true,
    special: "Deep dive"
  }
}
```

Now, everything is ready to add the **ngModel** directives to the form template. Take a look at the highlighted markup snippets in Listing 8-7.

Listing 8-7: dive-log-form.template.html (Exercise-08-02)

```
<form>
  <div class="row">
    <div class="col-sm-6">
      <div class="form-group">
        <label for="site">Dive site</label>
        <input class="form-control" id="site"
          [(ngModel)]="entry.site" name="site"
          placeholder="dive site">
      </div>
    </div>
    <div class="col-sm-6">
      <div class="form-group">
        <label for="location">Location</label>
        <input class="form-control" id="location"
          [(ngModel)]="entry.location" name="location"
          placeholder="location">
      </div>
    </div>
  </div>
  <div class="row">
    <div class="col-sm-3">
      <div class="form-group">
        <label for="depth">Max. depth</label>
        <input class="form-control" id="depth"
          [(ngModel)]="entry.depth" name="depth">
      </div>
    </div>
    <div class="col-sm-3">
      <div class="form-group">
        <label for="time">Bottom time</label>
        <input class="form-control" id="time"
          [(ngModel)]="entry.time" name="time">
      </div>
    </div>
  </div>
  <div class="checkbox">
    <label>
      <input type="checkbox"
        [(ngModel)]="entry.isFavorite" name="isFavorite">
      Mark as favorite
    </label>
  </div>
  <div class="form-group">
    <label>Special dive</label>
    <div class="radio"
      *ngFor="let diveType of specialDives">
```

```
      <label>
        <input type="radio" [value]="diveType"
        [(ngModel)]="entry.special" name="special">
        {{diveType}}
      </label>
    </div>
  </div>
  <div class="form-group">
    <label for="comments">Comments</label>
    <textarea class="form-control" id="comments" rows="5"
      [(ngModel)]="entry.coments" name="comments">
    </textarea>
  </div>
  <pre>{{entry | json}}</pre>
  <button class="btn btn-primary" type="submit">
    Submit
  </button>
</form>
```

The highlighted snippets contain two bindings, **[(ngModel)]**, and **name**, respectively. Both of them are important. We use **[(ngModel)]** with a template expression that defines the name of the property to which we want to bind the control. Behind the scenes, the form infrastructure keeps state information about each control. To keep this information separated from other control's status, each control must have a unique name—declared in the **name** attribute.

The first input control has this markup:

```
<input class="form-control" id="site"
  [(ngModel)]="entry.site" name="site"
  placeholder="dive site">
```

This declaration keeps the dive site information in the **entry.site** property of the form's component and the form infrastructure uses the "**site**" name internally to represent this control's state. The other controls apply the same approach to define their bindings.

The markup contains an additional entry to display the current model:

```
<pre>{{entry | json}}</pre>
```

While you are developing a form, such a simple diagnostic output may help you find issues. When you start the app, the form's controls display their initial values. As you change these values, you can immediately follow the changes in the diagnostic panel right above the submit button (Figure 8-3).

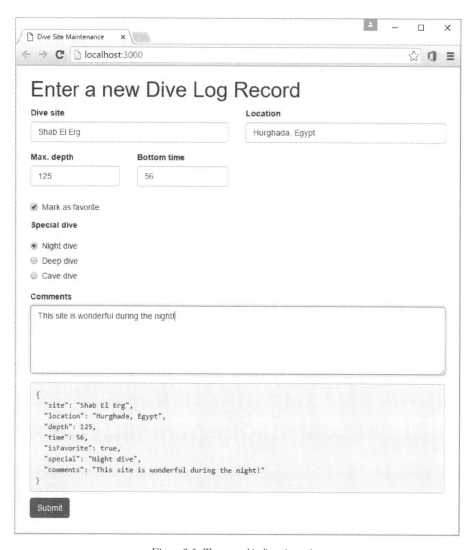

Figure 8-3: Two-way bindings in action

Property Names and ngModel

You should be careful when setting up **ngModel** bindings. If you specify a wrong property name or mistype one, your app still looks working properly—though it does not.

While I was preparing this sample, as mistyped the **comment** property, I wrote it with one "m" as "**coment**":

```
<textarea class="form-control" id="comments" rows="5"
  [(ngModel)]="entry.coments" name="comments">
</textarea>
```

Due to the dynamic nature of JavaScript, the framework put my comments into the **coments** property of **entry** within **DiveLogFormComponent**. I discovered this behavior only when I wrote code to display the submitted form information. That view always displayed an empty field, because the **comments** property used in the view's template was undefined—unlike **coments** with one "m".

With the help of the diagnostic output, I could easily catch this bug (Figure 8-4).

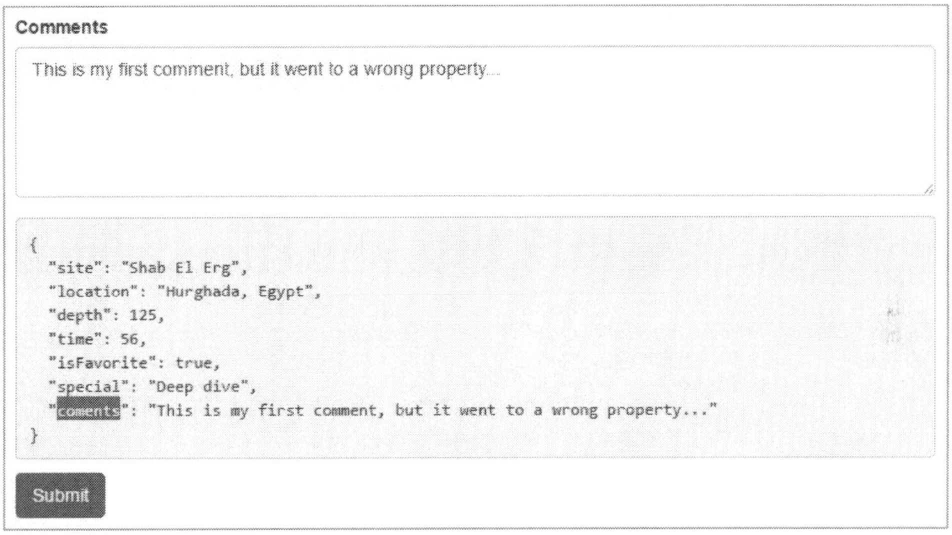

Figure 8-4: Diagnostic output help catching the typo

Banana in the Box

The Angular team and the community often mention the **[(ngModel)]** notation as "banana in the box". As soon as you omit **ngModel** from it, you will understand why: "**[()]**".

This notation is syntax sugar. Before processing such a binding, Angular transforms it. The **[(ngModel)]="*expr*"** binding is handled as if you wrote this:

```
[propName]="expr"
(exprChange)="expr=$event"
```

Angular turns the **[(ngModel)]="entry.site"** expression in the sample above to this construct:

```
<input [ngModel]="entry.site"
  (ngModelChange)="entry.site=$event" />
```

Theoretically, you can use the "banana in the box" syntax with your custom directives. So, if you had applied a **myDir** directive in a **<div>** tag, Angular would translate the **[(myDir)]="value"** binding to this one:

```
<div [myDir]="value" (myDirChange)="value=$event">
</div>
```

This transformation means that the **myDir** directive needs to handle the **myDirChange** event, too. I have not found any situation yet—except **ngModel**—when this syntax sugar would help me, nonetheless, you might meet one.

Multiple Controls Bound to the Same Property

Earlier I mentioned that each control must have a unique name. Well, it is not entirely correct. Sometimes we intend to bind the values coming from multiple controls to the same backing property. Take a look at this code snippet in the template definition:

```
<div class="radio"
  *ngFor="let diveType of specialDives">
  <label>
    <input type="radio" [value]="diveType"
    [(ngModel)]="entry.special" name="special">
    {{diveType}}
  </label>
</div>
```

The **ngFor** directive here generates three radio button controls bound to **entry.special**, with the internal name of "special". The form infrastructure understands our intention clearly. When we set the **entry.special** property in the code, only that radio button will be selected which has the value stored in **entry.special**. When the user selects a particular radio button, Angular stores the value of the selected button in **entry.special**.

Adding Basic Validation

The Angular forms infrastructure support several built-in HTML validators, such as **required**, **minlenght**, **maxlenght**, and **pattern**. When you add them to form controls, Angular validates their content accordingly. The **Exercise-08-03** folder contains a sample that adds several validators to the template (Listing 8-8).

Listing 8-8: dive-log-form.template.html (Exercise-08-03)

```
<form novalidate>
  <div class="row">
    <div class="col-sm-6">
      <div class="form-group">
        <label for="site">Dive site</label>
        <input class="form-control" id="site"
          [(ngModel)]="entry.site" name="site"
          required
          placeholder="dive site">
```

```html
            </div>
        </div>
        <div class="col-sm-6">
            <div class="form-group">
                <label for="location">Location</label>
                <input class="form-control" id="location"
                    [(ngModel)]="entry.location" name="location"
                    required
                    placeholder="location">
            </div>
        </div>
    </div>
    <div class="row">
        <div class="col-sm-3">
            <div class="form-group">
                <label for="depth">Max. depth</label>
                <input class="form-control" id="depth"
                    [(ngModel)]="entry.depth" name="depth"
                    required maxlength="3">
            </div>
        </div>
        <div class="col-sm-3">
            <div class="form-group">
                <label for="time">Bottom time</label>
                <input class="form-control" id="time"
                    [(ngModel)]="entry.time" name="time"
                    required maxlength="3">
            </div>
        </div>
        <!-- ... -->
    </div>
    <pre>{{entry | json}}</pre>
    <button class="btn btn-primary" type="submit">
        Submit
    </button>
</form>
```

As you can see from the code, the **site**, **location**, **depth**, and **time** properties are required. The **depth** and **time** values cannot be longer than three characters.

By default, the browser validates the form when we are about to submit it. Each browser vendor supports its own style to reflect validation errors, as demonstrated in Figure 8-5 and Figure 8-6.

This is not we would like to see. We want to provide our custom feedback, and as soon as possible, without waiting for clicking the Submit button. Moreover, we would like to disable the Submit button if the form has invalid data.

Figure 8-5: Invalid field value in Chrome

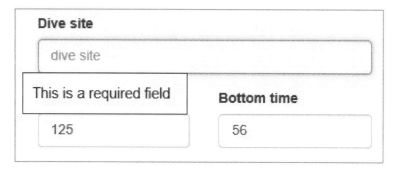

Figure 8-6: Invalid field value in Edge

Visual Validation Feedback

The **ngModel** directive tracks the state of controls through status properties. Table 8-1 summarizes them.

Table 8-1: ngModel status properties

Status Property	Description
touched	The control has been visited. This status is set right after the control loses the focus.
untouched	The control has not been visited yet. This status is set right after the control is initialized.
dirty	The control's value has changed. This status is immediately set as you start editing a control's value.
pristine	The control's value has not changed since it was initialized.

Status Property	Description
`valid`	The control's value is valid. This status is checked as you edit the control's value.
`invalid`	The control's value is invalid. This status is checked as you edit the control's value.

Each of these properties is a boolean flag that retrieves **true** if the corresponding control satisfies the condition.

Angular automatically decorates each control with CSS classes that represent the status flags. The names of these classes are **ng-touched**, **ng-untouched**, **ng-dirty**, **ng-pristine**, **ng-valid**, and **ng-invalid**, respectively. Logically, when the related status is true, the CSS class is added to the control's HTML element; otherwise, it is removed.

The **Exercise-08-04** folder contains a sample that demonstrates using these attributes. The component template in this sample is the same as in the previous exercise. The only difference is that **DiveLogFormComponent** applies CSS styles, as shown in Listing 8-9 and Listing 8-10.

Listing 8-9: dive-log-form.component.ts (Exercise-08-04)

```
import {Component} from '@angular/core';
import {DiveLogEntry} from './dive-log-entry';

@Component({
  selector: 'dive-log-form',
  templateUrl: 'app/dive-log-form.template.html',
  styleUrls: ['app/dive-log-form.styles.css']
})
export class DiveLogFormComponent {
  // ...
}
```

Listing 8-10: dive-log-form.style.css (Exercise-08-04)

```
.ng-valid[required] {
  border: 2px solid navy;
}

.ng-invalid:not(form) {
  border: 2px solid red;
}

.ng-dirty:not(form) {
  background-color: #e0e0e0;
}
```

```
.ng-pristine:not(form) {
  background-color: lightyellow;
}
```

You may wonder why the **.ng-invalid**, **.ng-dirty**, and **.ng-prisitine** style rules have the **:not(form)** selector tag. The answer is simple. The framework administers the status flags listed in Table 8-1 not only for the controls, but for the form itself. Omitting **:not(form)** would apply these style rules on the **<form>** HTML tag, too, and not just for its controls.

Figure 8-7 shows the app in action.

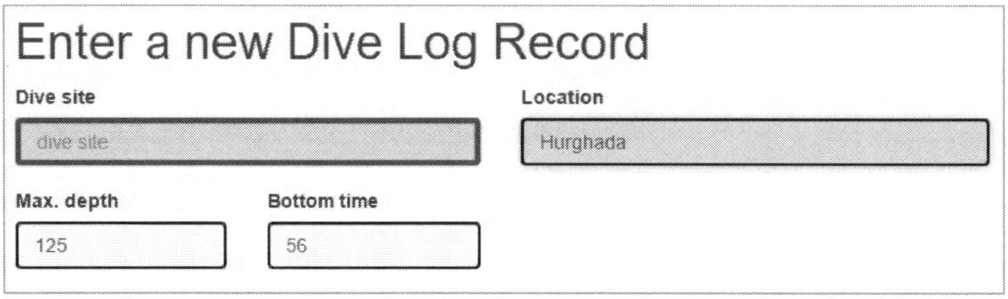

Figure 8-7: Style rules assigned to form control states

When we use a UI library, the design out-of-the-box often contains predefined CSS rules for the valid and invalid states of forms and controls. Such a library is Bootstrap, in which we can attach style classes like **has-error**, **has-success**, and **has-warning** to controls. To apply them, we need to access the status flags controls.

The sample in the **Exercise-08-05** folder demonstrates this situation. The sample does not use its own stylesheet; it leverages on Bootstrap style rules. Listing 8-11 shows the modifications of the component template.

Listing 8-11: dive-log-form.style.css (Exercise-08-05)

```
<form>
  <div class="row">
    <div class="col-sm-6">
      <div class="form-group"
        [class.has-error]="siteCtrl.invalid">
        <label for="site">Dive site</label>
        <input class="form-control" id="site"
          #siteCtrl="ngModel"
          [(ngModel)]="entry.site" name="site"
          required
          placeholder="dive site">
      </div>
    </div>
    <div class="col-sm-6">
      <div class="form-group"
        [class.has-error]="locationCtrl.invalid">
```

```
        <label for="location">Location</label>
        <input class="form-control" id="location"
          #locationCtrl="ngModel"
          [(ngModel)]="entry.location" name="location"
          required
          placeholder="location">
      </div>
    </div>
  </div>
  <div class="row">
    <div class="col-sm-3">
      <div class="form-group"
        [class.has-error]="depthCtrl.invalid">
        <label for="depth">Max. depth</label>
        <input class="form-control" id="depth"
          #depthCtrl="ngModel"
          [(ngModel)]="entry.depth" name="depth"
          required maxlength="3">
      </div>
    </div>
    <div class="col-sm-3">
      <div class="form-group"
        [class.has-error]="timeCtrl.invalid">
        <label for="time">Bottom time</label>
        <input class="form-control" id="time"
          #timeCtrl="ngModel"
          [(ngModel)]="entry.time" name="time"
          required maxlength="3">
      </div>
    </div>
  </div>
  <!-- .. -->
  <button class="btn btn-primary" type="submit">
    Submit
  </button>
</form>
```

The highlighted markup elements show something we have not met yet in the previous examples. There are template reference variables with template expressions like these:

```
#siteCtrl="ngModel"
#timeCtrl="ngModel"
```

Why do we assign the "**ngModel**" template expression value to the template reference variable?

By default, when we add a template reference variable—such as **#siteCtrl**, or **#timeCtrl**—, those variables get a reference to their host element. However, in many cases, we are rather interested in getting a reference to a directive assigned to the host element. We can access these directives through the expression assigned to the template reference variable.

The framework adds a reference to a directive in its host element's DOM representation. When the directive is declared, we can set a name to assign the directive to a variable. For example, the **NgModel** directive has this metadata:

```
@Directive({
    selector: '[ngModel]:not([formControlName]):not([formControl])',
    providers: [formControlBinding],
    exportAs: 'ngModel'
})
```

The **exportAs** metadata property defines the name we can use to reference to this directive. Thus, when we say **#siteCtrl="ngModel"**, we get a template reference to the **ngModel** directive associated with the **<input>** tag. Now, it is easier to understand how this markup works:

```
<div class="form-group"
    [class.has-error]="siteCtrl.invalid">
    <label for="site">Dive site</label>
    <input class="form-control" id="site"
        #siteCtrl="ngModel"
        [(ngModel)]="entry.site" name="site"
        required
        placeholder="dive site">
</div>
```

The **siteCtrl** template reference variable points to the **ngModel** directive instance that holds the control's status flags. If the **invalid** flag of the **ngModel** we access through **siteCtrl** evaluates to true, the **has-error** style class is added to the classes of the **form-group <div>** tag.

When you run the app, you can check that invalid fields apply the **has-error** Bootstrap style (Figure 8-8).

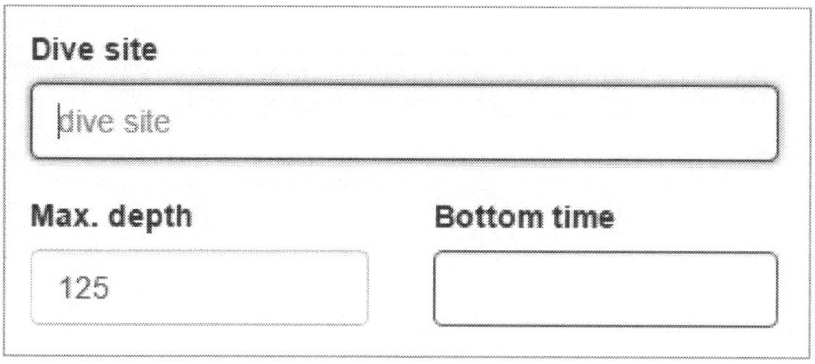

Figure 8-8: Bootstrap styles applied

Adding Validation Messages

Our users will not be happy about having an invalid field without any further explanation. We can easily add validation messages using the status flags—and other control status information. The

Exercise-08-06 folder contains a modified sample that uses the markup snippet shown in Listing 8-12.

Listing 8-12: dive-log-form.template.html (Exercise-08-06)

```
...
<div class="form-group"
  [class.has-error]="siteCtrl.invalid">
  <label for="site">Dive site</label>
  <input class="form-control" id="site"
    #siteCtrl="ngModel"
    [(ngModel)]="entry.site" name="site"
    required
    pattern="[A-Za-z]*"
    placeholder="dive site">
  <span *ngIf="siteCtrl.errors?.required" class="help-block">
    Please specify a site name.
  </span>
  <span *ngIf="siteCtrl.errors?.pattern" class="help-block">
    Please use only letter characters.
  </span>
</div>
...
```

This code snippet assigns two validators, **required** and **pattern** to the **<input>** tag. The **pattern** value specifies that only letters should be accepted. The **siteCtrl** reference points to the **ngModel** of the **<input>** element. Beside the status flags and other state information, we can access an **errors** object that provides more detailed information about validation issues. As the two **<span>** tags demonstrate, we can make a distinction' between the validation errors caused by the **required** and **pattern** validators.

> *NOTE: If the control value is valid, the **errors** property evaluates to **undefined**, so we apply the Elvis operator (?.) to access the **required** and **pattern** properties.*

Figure 8-9 and Figure 8-10 depict that the two different issue results in two different validation message.

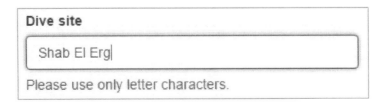

Figure 8-9: Message from the required validator

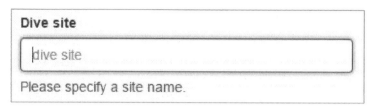

Figure 8-10: Message from the pattern validator

Submitting a Form

After filling the form in, the user wants to submit it. In a single page application, we do not intend to follow the traditional pattern to post back the form to the server. We rather process the data at the browser.

The sample in the **Exercise-08-07** folder allows the user to submit the form. Because we would like to display the submitted data, we modify the **DiveLogFormComponent** class, as shown in Listing 8-13.

Listing 8-13: dive-log-form.component.ts (Exercise-08-07)

```
...
export class DiveLogFormComponent {
  // ...

  submitted = false;

  submitForm() {
    this.submitted = true;
  }
}
```

When the user submits the form, we invoke the **submitForm()** method—and it sets the **submitted** flag.

With a few small changes in the component template, we complete data submission (Listing 8-14).

Listing 8-14: dive-log-form.template.html (Exercise-08-07)

```
<form *ngIf="!submitted"
  novalidate
  (ngSubmit)="submitForm()"
  #entryForm="ngForm">

  <!-- Unchanged, omitted for the sake of brevity -->
```

```
  <button class="btn btn-primary" type="submit"
    [disabled]="entryForm.invalid">
    Submit
  </button>
</form>
<div *ngIf="submitted">
  <h2>The form has been submitted with this data:</h2>
  <pre>{{entry | json}}</pre>
</div>
```

In the **<form>** tag, we create a template reference variable (**entryForm**) to the **NgForm** directive, which represents the state of the form. In the template expression of **entryForm**, we use the "**ngForm**" expression, since the directive is exported with this name.

Just as individual controls, the **NgForm** directive has the status flags in Table 8-1, too. With the test of **entryForm.invalid**, we can disable the Submit button given there is any error within the form.

The **novalidate** standard HTML attribute prevents the browser to validate the form when the submit command is executed, and thus the form can be submitted even with invalid data. Here, we do our own validation, so it is indifferent whether we add **novalidate** or not. I suggest you to add it to **<form>** when you carry out custom validation.

Whenever the framework submits the form, it raises the **ngSubmit** event. In response, we invoke the **submitForm()** method and change the **submitted** flag. The framework detects this change, removes the form and displays the submitted data (Figure 8-11).

Figure 8-11: The submitted form data

Reactive Forms

The template-driven model you get acquainted with the previous section was simple-to-use. You just created a form template with controls, validation attributes, and directives. During run time, Angular interpreted the template and represented it with a control model. The framework assigned an **NgForm** directive to the **<form>** element, and **NgModel** instances to each control with an **ngModel** binding.

Angular supports another approach, Reactive Forms. You still need to create a form template with this method, but this template is simpler and shorter. Instead applying validation attributes and directives to the template, you write the control model in code.

This way of form management requires more effort than the template-driven approach, but we gain flexibility. We can easily create dynamic forms—for example, a survey using questions from a database, or change the validation model on the fly.

In this section, we move the previous sample to use the Reactive Form approach.

Moving to ReactiveFormsModule

Besides **FormsModule**, Angular ships another one, **ReactiveFormsModule**. If you decide to apply this latter approach, you have to import **ReactiveFromsModule** into your app, as shown in Listing 8-15.

Listing 8-15: app.module.ts (Exercise-08-08)

```
import {NgModule} from '@angular/core';
import {BrowserModule} from '@angular/platform-browser';
import {ReactiveFormsModule} from '@angular/forms';

import {AppComponent} from './app.component';
import {DiveLogFormComponent} from './dive-log-form.component'

@NgModule({
  imports: [
    BrowserModule,
    ReactiveFormsModule
  ],
  declarations: [
    AppComponent,
    DiveLogFormComponent
  ],
  bootstrap: [AppComponent]
})
export class AppModule { }
```

Changing the Form Template

To create our form control model, we need to change the **DiveLogComponentForm** template. We remove the **ngModel** bindings, the validation attributes, and the template reference variables. To support the control model, we add a few new directives to the template. Take a look at Listing 8-16 that highlights the key changes.

Listing 8-16: dive-log-form.template.html (Exercise-08-08)

```
<form *ngIf="!submitted" (ngSubmit)="submitForm()"
  [formGroup]="diveLogForm" >
  <div class="row">
    <div class="col-sm-6">
      <div class="form-group"
        [class.has-error]="isInvalid('site')">
        <label for="site">Dive site</label>
        <input class="form-control" id="site"
          formControlName="site"
          placeholder="dive site">
        <span *ngIf="isInvalid('site')" class="help-block">
          Please specify a site name.
        </span>
      </div>
    </div>
    <div class="col-sm-6">
      <div class="form-group"
        [class.has-error]="isInvalid('location')">
        <label for="location">Location</label>
        <input class="form-control" id="location"
          formControlName="location"
          placeholder="location">
        <span *ngIf="isInvalid('location')" class="help-block">
          Please specify a site name.
        </span>
      </div>
    </div>
  </div>
  <div class="row">
    <div class="col-sm-3">
      <div class="form-group"
        [class.has-error]="isInvalid('depth')">
        <label for="depth">Max. depth</label>
        <input class="form-control" id="depth"
          formControlName="depth">
        <span *ngIf="isInvalid('depth')" class="help-block">
          Please specify a valid depth.
        </span>
      </div>
    </div>
    <div class="col-sm-3">
      <div class="form-group"
```

```
      [class.has-error]="isInvalid('time')">
      <label for="time">Bottom time</label>
      <input class="form-control" id="time"
        formControlName="time">
      <span *ngIf="isInvalid('time')" class="help-block">
        Please specify a valid time.
      </span>
    </div>
  </div>
</div>
<div class="checkbox">
  <label>
    <input type="checkbox"
      formControlName="isFavorite">
    Mark as favorite
  </label>
</div>
<div class="form-group">
  <label>Special dive</label>
  <div class="radio"
    *ngFor="let diveType of specialDives">
    <label>
      <input type="radio" [value]="diveType"
      formControlName="special" >
      {{diveType}}
    </label>
  </div>
</div>
<div class="form-group">
  <label for="comments">Comments</label>
  <textarea class="form-control" id="comments" rows="5"
    formControlName="comments">
  </textarea>
</div>
<button class="btn btn-primary" type="submit"
  [disabled]="diveLogForm.invalid">
  Submit
</button>
</form>
<div *ngIf="submitted">
  <h2>The form has been submitted with this data:</h2>
  <pre>{{submittedDive | json}}</pre>
</div>
```

The first—and most important—change is that we add the **formGroup** directive to the **<form>** element. In the code, Reactive Forms represents the abstraction of the form—we can say, its model—with a **FormGroup** instance, which aggregates its child controls' state. For example, when any of the **FormGroup** children is invalid, the **FormGroup** becomes invalid, too.

NOTE: *We can nest* **FormGroup** *instances into each other. Thus we can create hierarchical forms with Angular. This is why* **FormGroup** *is called* **FormGroup**, *and not* **Form**.

With the **[formGroup]="diveLogForm"** binding we declare that we create the **FormGroup** instance in the component class and make it available through the **diveLogForm** property.

NOTE: *The* **<div>** *tag with the* **form-group** *style class has nothing to do with the* **formGroup** *directive. The* **form-group** *style is defined in Bootstrap.*

The second change is that each control now applies the **formControlName** directive. This directive's value is the glue that ties a particular child control of **FormGroup** with its HTML representation. **formControlName** allows us to exile the **name** attribute.

The third change is that we invoke the **isInvalid()** method of the component class to check whether a certain control is invalid. We pass the name of the control to **IsInvalid()**—the same value we use in **formControlName**.

With these changes, we have shrunk the template definition of a single control. For example, the **site** control's definition was this with the template-driven approach:

```
<div class="form-group"
  [class.has-error]="siteCtrl.invalid">
  <label for="site">Dive site</label>
  <input class="form-control" id="site"
    #siteCtrl="ngModel"
    [(ngModel)]="entry.site" name="site"
    required
    placeholder="dive site">
  <span *ngIf="siteCtrl.invalid" class="help-block">
    Please specify a site name.
  </span>
</div>
```

After the change, it became leaner:

```
<div class="form-group"
  [class.has-error]="isInvalid('site')">
  <label for="site">Dive site</label>
  <input class="form-control" id="site"
    formControlName="site"
    placeholder="dive site">
  <span *ngIf="isInvalid('site')" class="help-block">
    Please specify a site name.
  </span>
</div>
```

Changing the Component Class

Though we have a new template, we have not created the control model that works with the form. As Listing 8-17 shows, we create it in the component class.

Listing 8-17: dive-log-form.component.ts (Exercise-08-08)

```
import {Component, Input, OnInit} from '@angular/core';
import {FormGroup, FormControl, Validators} from '@angular/forms'
import {DiveLogEntry} from './dive-log-entry';

@Component({
  selector: 'dive-log-form',
  templateUrl: 'app/dive-log-form.template.html'
})
export class DiveLogFormComponent implements OnInit {
  specialDives = [
    "Night dive",
    "Deep dive",
    "Cave dive"
  ]

  entry: DiveLogEntry = {
    site: 'Shab El Erg',
    location: 'Hurghada, Egypt',
    depth: 125,
    time: 56,
    isFavorite: true,
    special: "Deep dive"
  }

  @Input() diveLogForm: FormGroup;

  ngOnInit() {
    this.diveLogForm = new FormGroup({
      site: new FormControl(this.entry.site,
        Validators.required),
      location: new FormControl(this.entry.location,
        Validators.required),
      depth: new FormControl(this.entry.depth,
        Validators.required),
      time: new FormControl(this.entry.time,
        Validators.required),
      isFavorite: new FormControl(this.entry.isFavorite),
      special: new FormControl(this.entry.special),
      comments: new FormControl(this.entry.comments),
    });
  }
```

```
isInvalid(controlName: string) {
  return this.diveLogForm.controls[controlName].invalid
}

submitted = false;
submittedDive: DiveLogEntry;

submitForm() {
  this.submittedDive = this.diveLogForm.value;
  this.submitted = true;
}
}
```

The component class utilizes the **OnInit** lifecycle hook to set up the **diveLogForm** property that describes the control model of the form. The **FromGroup** constructor accepts an object where each property represents a particular control. The control names must match with the ones we specified in the **formControlName** directive values earlier; otherwise, the framework cannot bind the model to the corresponding HTML controls.

Property values are **FormControl** instances. We pass the initial value of the control and a set of optional **Validators** to each **FormControl** constructor.

> **NOTE**: *In this very case, we could have initialized **diveLogForm** in the constructor. However, this is not a good practice. In most cases, the control model may utilize input property values that are not initialized yet in the constructor. Putting this login into **ngOnInit()** helps you avoid such issues.*

The **FormGroup** instance stored in **diveLogForm** makes it easy to get state information about a particular control. Its **controls** property allows checking—among the others— the status flags summarized in Table 8-1. The code of **isInvalid()** is straightforward:

```
isInvalid(controlName: string) {
  return this.diveLogForm.controls[controlName].invalid
}
```

There is a crucial difference between the template-driven approach and Reactive Forms. While the template-driven method uses two-way bindings with **[(ngModel)]**, Reactive Forms bindings are unidirectional. Although we used the **entry** property of the component class to set up the initial control values, the changes are never written back to **entry**. The code of **submitForm()** shows how data submission works:

```
submitted = false;
submittedDive: DiveLogEntry;

submitForm() {
  this.submittedDive = this.diveLogForm.value;
  this.submitted = true;
}
```

At the moment of clicking Submit, we save the current state of the form into **submittedDive**. Due to the **value** property of **FormGroup**, we can make a simple assignment. Observe, we ensure that **submittedDive** and **diveLogForm.value** have the same shape by using the same names in the control properties and **DiveLogEntry**.

To display the submitted set of data, we modify the template part to display **submittedDive**:

```
...
<div *ngIf="submitted">
  <h2>The form has been submitted with this data:</h2>
  <pre>{{submittedDive | json}}</pre>
</div>
...
```

When you run the app, you can check that it works just like the previous one—however, this time with the Reactive Forms approach.

Using a Form Builder

The syntax we used within **ngOnInit()** to create the **FormGroup** and its **FormControl** children is verbose. Reactive Forms provides a **FormBuilder** object that allows creating the control model in a shorter and more readable way, as shown in Listing 8-18.

Listing 8-18: dive-log-form.component.ts (Exercise-08-09)

```
import {Component, Input, OnInit} from '@angular/core';
import {FormGroup, FormControl, Validators} from '@angular/forms'
import {FormBuilder} from '@angular/forms'
import {DiveLogEntry} from './dive-log-entry';

@Component({
  selector: 'dive-log-form',
  templateUrl: 'app/dive-log-form.template.html'
})
export class DiveLogFormComponent implements OnInit {
  // ...
  constructor (private builder: FormBuilder) { }

  @Input() diveLogForm: FormGroup;

    ngOnInit() {
    this.diveLogForm = this.builder.group({
      site: [this.entry.site, Validators.required],
      location: [this.entry.location, Validators.required],
      depth: [this.entry.depth, Validators.required],
      time: [this.entry.depth, Validators.required],
```

```
      isFavorite: [this.entry.isFavorite],
      special: [this.entry.special],
      comments: [this.entry.comments]
    });
  }
  // ...
}
```

We obtain a **FormBuilder** object through the constructor. In the body of **ngOnInit()**, we pass a build configuration object to the **group()** method of the builder.

Custom Validation

The built-in validators are very useful, but scarcely enough. We often need to check control validity according to custom rules. With Reactive Forms, it is easy to write custom validation.

Let's extend the previous sample so that we can check the depth and time values. We can accept only numeric values; the depth must be between 0 and 130 feet; while the time between 0 and 240 minutes.

We can write our custom validation function and pass them to the control model. A validation function has this signature:

```
validator(control: AbstractControl): {[key:string]: any}
```

It means that the **validator** function accepts an object that represents the control. It must return an object that describes the validation error. If the control's value is valid, the function should return null.

Listing 8-19 shows the validator that checks the **depth** value. Observe, it carries out a couple of tests.

Listing 8-19: dive-log-form.component.ts (Exercise-08-10)

```
...
depthValidator(control: AbstractControl): {[key:string]: any} {
  let value = control.value;
  if (!value) return null;
  if (isNaN(value)) {
    return { NaN: true };
  } else {
    let depth = parseInt(value, 10);
    if (depth >= 0 && depth <= 130) {
      return null;
    }
  }
  return { depth: {min: 0, max: 130} };
}
...
```

After reading out the control's value, **depthValidator()** returns immediately with success (**null** value), provided the control's value is unspecified. It is not the task of this validator to catch such an issue—**Validators.required** does this. When the control has a value, the method returns different error objects if the value is not numeric and if it is out of the accepted range. Observe, when it reports a range error, it retrieves the expected boundaries in the error object for future use.

The other validator function, **timeValidator()**, has the very same structure. Listing 8-20 shows how we can assign these validators to the control model.

Listing 8-20: dive-log-form.component.ts (Exercise-08-10)

```
...
ngOnInit() {
  this.diveLogForm = this.builder.group({
    site: [this.entry.site, Validators.required],
    location: [this.entry.location, Validators.required],
    depth: [this.entry.depth,
      [Validators.required, this.depthValidator]],
    time: [this.entry.depth,
      [Validators.required, this.timeValidator]],
    isFavorite: [this.entry.isFavorite],
    special: [this.entry.special],
    comments: [this.entry.comments]
  });
}
...
```

Besides introducing custom validation, we remove the error message literals from the template and move them to the component class. **DiveLogFormComponent** defines the **getValidationMessage()** method (Listing 8-21), which relies on the message literals stored in the **validationMessages** object.

Listing 8-21: dive-log-form.component.ts (Exercise-08-10)

```
...
getValidationMessage(controlName: string) {
  let message = '';
  let control = this.diveLogForm.get(controlName);
  if (control) {
    let messages = this.validationMessages[controlName];
    if (messages && control.errors) {
      for (const key in control.errors) {
        message += messages[key] + ' ';
      }
    }
  }
  return message == ''
    ? 'Control value is invalid.'
```

```
    : message;
  }

validationMessages = {
  site: {
    required: 'Please specify a site name',
  },
  location: {
    required: 'Please specify a location',
  },
  depth: {
    required: 'Please specify a depth',
    NaN: 'Value must be a number',
    depth: 'Depth must be between 0 and 130'
  },
  time: {
    required: 'Please specify a time',
    NaN: 'Value must be a number',
    time: 'Time must be between 0 and 240'
  },
};
...
```

To obtain the current state of a particular control, **getValidationMessage()** invokes the **get()** method of **diveLogForm** with the name of the control. If that is invalid, its **errors** property contains an object with its key names enumerating the types of validation issues.

In Listing 8-19, the **depthValidator()** returned two different error object, one with a **NaN** property, and another with **depth** property. These error objects are appended to the **errors** property of **depth**, so when the corresponding validation issue occurs, **validationMessages** contains an appropriate message.

To apply **getValidationMessage()**, we need to modify the error **<span>** of each validated control, as shown in the example of the site control:

```
<span *ngIf="isInvalid('site')" class="help-block">
  {{getValidationMessage('site')}}
</span>
```

When you run the app, the custom validation works as expected (Figure 8-12).

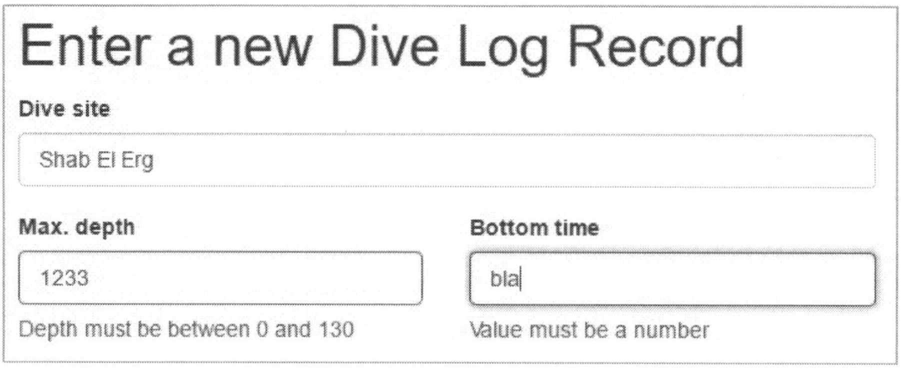

Enter a new Dive Log Record

Dive site

Shab El Erg

Max. depth

1233

Depth must be between 0 and 130

Bottom time

bla

Value must be a number

Figure 8-12: Custom validation in action

Summary

Angular delivers two modules, `FormsModule` and `ReactiveFormsModule` to manage forms. They provide different approaches.

`FormsModule` allows creating forms with the template-driven approach. You create a template, assign validation attributes and directives to it to declare your intentions to control the form. When you run the app, Angular interprets the template and derives its form control model. This method utilizes two-way bindings.

With `ReactiveFormsModule`, you can define forms with the Reactive Forms approach—using unidirectional bindings. You still create a template for the form, but you set up the control model in code. Though it requires a bit more effort than the template-driven approach, it provides better flexibility.

Both models help you manage the state of the form including change tracking, validation, and data submission.

Chapter 9: The Component Router

WHAT YOU WILL LEARN IN THIS CHAPTER

Understanding how you can use the Component Router

Creating routing arrays and configuring routing

Setting up hierarchical routing

Getting acquainted with routing guards

Configuring modules for asynchronous loading

In *Chapter 3, Routing*, we already created a sample application (dive site management) that leveraged the Angular Component router. That time, we just scratched the surface of what we can do with this excellent feature. In this chapter, we dive deeper into Angular routing.

Understanding Angular Routing

The main feature of the Angular Component Router is that it understands URLs we enter in the address bar and navigates to a component associated with that URL. The target component can access the parameters and options passed in the address, process them and render the view accordingly.

We can navigate with HTML links just as we do in a plain HTML application. Also, we can move from one URL to another within the page programmatically, for example, as a response to clicking a button, selecting an item from a list, choosing a new radio button, and so on. As the code changes the current navigation route, Angular updates the URL in the address bar.

Let's get acquainted with a few essential things that help us understand how routing works.

Declarative Routing

Angular provides declarative routing through an array of **Route** objects. For example, if we have three components, **PearComponent**, **OrangeComponent**, and **LemonComponent**, we can set up the routing with this array:

```
routes: Routes = [
  { path: 'pear', component: PearComponent },
  { path: 'orange', component: OrangeComponent },
  { path: 'lemon', component: LemonComponent },
  { path: '', pathMatch: 'full', component: OrangeComponent }
```

```
];
```

Provided our app is available on **http://myfruits.com**, the Component Router directs the **http://myfruits/pear**, **http://myfruits/orange**, and **http://myfruits/lemon** URLs to **PearComponent**, **OrangeComponent**, and **LemonComponent**, respectively. The last item of routes ensures that the root URL—if we navigate to **http://myfruits.com**—activates **OrangeComponent**.

Host View, Routed View, and Router Outlet

When we bootstrap the app, we can pass the array of the routes to the root application context. The bootstrapped component's view becomes the *host view*. As the router navigates to a URL, it displays the target component's view—the *routed view*—within the host view. The host component must provide an area within its template that designates the patch of UI into which we intend to display the routed view. This area is the *router outlet*. In the template of the host, we use the **<router-outlet>** tag—the selector of the **RouterOutlet** directive—to sign the location for the routed view. Let's assume that the host component of **http://myfruits.com** has this template:

```
<!-- Menu ... -->
<!-- Sidebar ... -->
<div class="container">
  <h2>Welcome to MyFruits.com</h1>
  <!-- Main panel -->
  <router-outlet></router-outlet>
</div>
```

When the user navigates to **http://myfruits.com/lemon**, the main panel displays the view of **LemonComponent** (Figure 9-1).

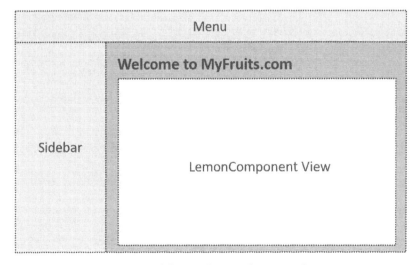

Figure 9-1: The routed view in the router outlet

Hierarchical Roots

Angular allows routed views to be host views at the same time. They can provide their child routing information and set a router outlet in their templates. For example, `LemonComponent` may have this route array:

```
lemonRoutes: Routes = [
   { path: 'green', component: GreenLemonComponent },
   { path: 'red', component: RedLemonComponent },
   { path: '', component: DefaultLemonComponent }
];
```

If the user—or the app—navigates to the `http://myfruits.com/lemon/red` URL, the Component Router utilizes the "lemon" part of the address to render `LemonComponent` in the main panel's router outlet. The router applies the "red" part to display the view of `RedLemonComponent` within the router outlet of `LemonComponent` (Figure 9-2).

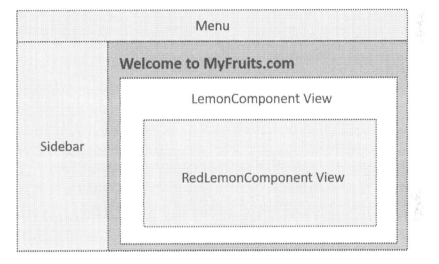

Figure 9-2: Nested routes and views

Route Parameters

The Component Router accepts parameter declarations in route definitions, and the target component can access the current parameter value. We can change the `lemonRoutes` array so that the "red" option has an identifier as a parameter:

```
lemonRoutes: Routes = [
   { path: 'green', component: GreenLemonComponent },
   { path: 'red/:id', component: RedLemonComponent },
   { path: '', component: DefaultLemonComponent }
];
```

When the app navigates to the **http://myfruits.com/lemon/red/123** URL, the target **RedLemonComponent** can access the current value of the **id** parameter (123), and utilize it to render the component's view (Figure 9-3).

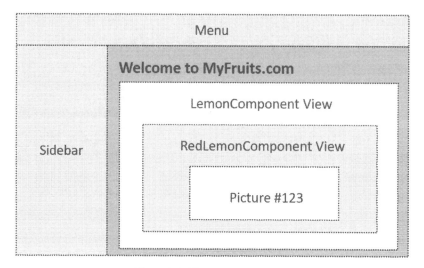

Figure 9-3: Using a route parameter

Guards

There are scenarios when we need to prevent navigating to a particular route. For example, if we need administrative permissions to access the **http://myfruits.com/lemon/admin** route, we need to check this privilege before navigation. In other cases, when we typed data in a web form, we want to prevent a sudden navigation away from the current view without confirmation from the user.

The Component Router allows associating *guards* to routes. With these guards, we can prevent that any user can navigate anywhere in the application anytime.

Routing in Action

In this chapter, we are going to build up a simple application that allows us to get acquainted with the Angular Component Router. This sample app provides a welcome page, allows viewing dive log information, editing dive site information, and manages the privileges of anonymous users, logged-in users, and administrators.

Setting Up Routing

Let's start with setting up the blueprint of the application from the pieces we already created in the previous chapters of the book. When this setup is ready, we will have an app with three routes that are managed by `WelcomeComponent` (Figure 9-4), `DiveLogComponent` (Figure 9-5), and `SiteListComponent` (Figure 9-6), respectively.

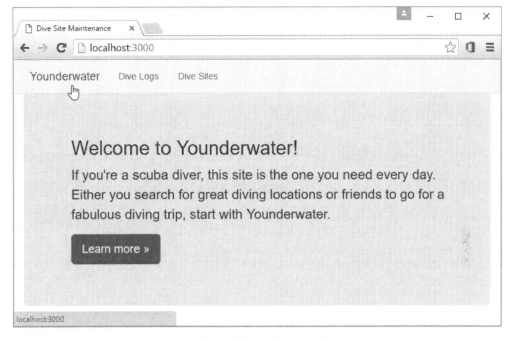

Figure 9-4: The welcome panel

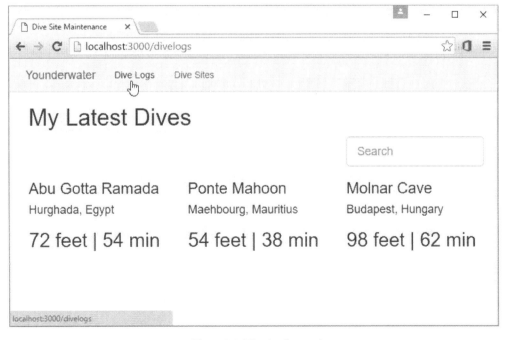

Figure 9-5: The dive log panel

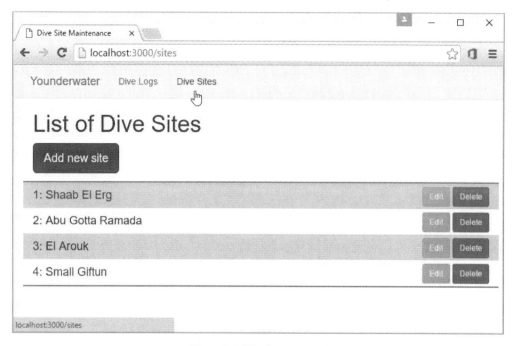

Figure 9-6: The dive sites panel

I re-used the **DiveLogComponent** from Chapter 1 and the **DiveSiteComponent** from Chapter 3. For the sake of this demonstration, I slightly modified them: encapsulated both of these components to their dedicated feature modules and put them into separate folders, **logs**, and **sites**, respectively—while keeping their original operation intact.

WelcomeComponent is a new element; its source files are in the **welcome** folder. As you can see in Listing 9-1, Listing 9-2, and Listing 9-3, the corresponding code is very simple.

Listing 9-1: welcome.component.ts (Exercise-09-01)

```
import {Component} from '@angular/core';

@Component({
  selector: 'welcome',
  templateUrl: 'app/welcome/welcome.template.html'
})
export class WelcomeComponent {
}
```

Listing 9-2: welcome.template.html (Exercise-09-01)

```
<div class="jumbotron">
  <div class="container">
    <h2>Welcome to Younderwater!</h2>
    <p>
      If you're a scuba diver, this site is the
```

```
        one you need every day. Either you search
        ...
    </p>
    <p><a class="btn btn-primary btn-lg" role="button">
      Learn more &raquo;
    </a></p>
  </div>
</div>
```

Listing 9-3: welcome.module.ts (Exercise-09-01)

```
import {NgModule} from '@angular/core';
import {BrowserModule} from '@angular/platform-browser';
import {WelcomeComponent} from './welcome.component';

@NgModule({
  imports: [BrowserModule],
  declarations: [WelcomeComponent]
})
export class WelcomeModule { }
```

To support routing, we need to add an HTML **<base>** element to **index.html**. This tag tells the router how to compose the navigation URLs (Listing 9-4).

Listing 9-4: index.html (Exercise-09-01)

```
<html>
<head>
  <base href="/">
  <title>Dive Site Maintenance</title>
  <link href="/node modules/bootstrap/dist/css/bootstrap.min.css"
    rel="stylesheet" />
  <script src="/node_modules/jquery/dist/jquery.min.js"></script>
  <!-- ... -->
</head>
  <!-- ... -->
</html>
```

Without **<base>** the router will raise an error message.

> *NOTE: There are programmatic ways to add the **<base>** tag from code, or inject this information into the router. In this chapter, we will use the **<base>** tag.*

The router needs an array of routes so that it can navigate to components. By convention, we put this information into a separate file—**app.routes.ts**, as shown in Listing 9-5.

Listing 9-5: app.routes.ts (Exercise-09-01)

```
import {Routes, RouterModule} from '@angular/router';
import {WelcomeComponent} from './welcome/welcome.component';
import {DiveLogComponent} from './logs/dive-log.component';
import {SiteListComponent} from './sites/site-list.component';

const routes: Routes = [
  { path: 'divelogs', component: DiveLogComponent },
  { path: 'sites', component: SiteListComponent },
  { path: '', pathMatch: 'full', component: WelcomeComponent }
];

export const routingModule = RouterModule.forRoot(routes);
```

This code consumes two essential types. **Routes** is an array of **Route** instances. **Route** describes an object in the routing table. The Component Router applies this information to choose the component responsible for rendering the view of a particular route.

RouterModule represents an Angular module that provides routing information for the app. With the **RouterModule.forRoot(routes)** call we create a module that contains the routing data for our app. We store this value in the exported **routingModule** variable so that we can import it in the application root module.

Let's see, how we tell the app to use the information in **app.routes.ts** should be utilized. Listing 9-6 unravels the secret.

Listing 9-6: app.routes.ts (Exercise-09-01)

```
import {NgModule} from '@angular/core';
import {BrowserModule} from '@angular/platform-browser';
import {AppComponent} from './app.component';

import {WelcomeModule} from './welcome/welcome.module';
import {DiveLogModule} from './logs/dive-log.module';
import {SitesModule} from './sites/sites.module';

import {routingModule} from './app.routes';

@NgModule({
  imports: [
    BrowserModule,
    WelcomeModule,
    DiveLogModule,
    SitesModule,
    routingModule
  ],
```

```
    declarations: [
      AppComponent
    ],
    bootstrap: [AppComponent]
})
export class AppModule { }
```

Without the highlighted code, the application root module looks as usual. We import **BrowserModule**—as a core requirement for every Angular app running in the browser—, and the three feature modules, **WelcomeModule**, **DiveLogModule**, and **SitesModule**, respectively. We declare only **AppComponent** and use it as the root component to bootstrap the app.

The highlighted lines import a new module, **routingModule**, into the app. Why do we load a module to configure routing?

In *Chapter 6, Understanding NgRoutes*, you learned that the framework uses the imported modules to look for directives, services, components, pipes, and other objects to resolve references. For example, this is the way Angular finds the **ngIf** and **ngFor** directives in **BrowserModule**.

The same mechanism works when we import **routingModule**. This module is an instance of the **ModuleWithProviders** class and encapsulates those objects—directives, services, and others—that the app will use for routing. Most importantly, **routerModule** offers services configured according to the routing data we specified in **app.routing.ts**. Whenever the framework—or our app—needs to use such a service, we can be sure that that is set up correctly.

Listing 9-7 shows the template of **AppComponent**.

Listing 9-7: app.template.html (Exercise-09-01)

```html
<nav class="navbar navbar-default navbar-fixed-top" role="navigation">
  <div class="container">
    <button type="button" class="navbar-toggle collapsed"
            data-toggle="collapse"
            data-target="#adminMenu">
      <span class="sr-only">Toggle navigation</span>
      <span class="icon-bar"></span>
      <span class="icon-bar"></span>
      <span class="icon-bar"></span>
    </button>
    <a class="navbar-brand" routerLink="/">Younderwater</a>
    <div class="collapse navbar-collapse" id="adminMenu">
      <ul class="nav navbar-nav">
        <li>
          <a routerLink="/divelogs" routerLinkActive="active">
            Dive Logs
          </a>
        </li>
        <li>
          <a routerLink="/sites" routerLinkActive="active">
            Dive Sites
```

```
            </a>
         </li>
      </ul>
    </div>
  </div>
</nav>
<div class="container-fluid">
  <router-outlet></router-outlet>
</div>
```

The majority of this markup is used by the Bootstrap component to render the menu in the page. From routing point of view, we need to focus on the four highlighted line. The most important of them is the `<router-outlet>` tag. If we left this, the router would raise an error, because it would not find an outlet to load the routed view.

The other three lines represent links that navigate to the corresponding components. Instead of using the `href` attribute with the anchor tag, we can use the `routerLink` directive. Here, we use `routerLink` with a string literal, but we can use it with a template expression, too. The directive composes a link from its value—the value of the template expression—and creates an `href` attribute accordingly.

The `routerLinkActive` directive offers a useful feature. When the current route becomes the one represented by the anchor tag, `routerLinkActive` toggles the CSS class we specify in its value. So when "/divelog" becomes the active root, the related anchor element gets the "active" class; otherwise, this class is removed from the element. In the code above, we use the "active" class— provided by Bootstrap—to highlight the active menu link.

Start the app. When the page is loaded, the app displays `WelcomeComponent`., according to the highlighted route definition in `app.routes.ts`:

```
const routes: Routes = [
  { path: 'divelogs', component: DiveLogComponent },
  { path: 'sites', component: SiteListComponent },
  { path: '', pathMatch: 'full', component: WelcomeComponent }
];
```

Because the empty path matches with any URL, setting `pathMatch` to "full" ensures that the router will navigate to `WelcomeComponent` if and only if we specify the base URL.

Play with the app and test how clicking the menu items navigates in the app. Check that as you move among the links, the address bar is updated accordingly. Try specifying valid URL routes in the browser directly, and see how they navigate. Observe the error message in the console output when you try to apply an unsupported URL within the app.

Configuring Child Routes

In Chapter 3, the dive site management app leveraged the Component Router. Nonetheless, in **Exercise-09-01**, we created a new routing table, and that prevents the Dive Sites panel from working correctly. When you click the Add new site button, you get a "*Cannot match any routes: 'add'*" error message. The button tries to navigate to the **/add** URL, but the router understands only the base URL, **/divelogs**, and **/sites**.

The dive site management feature is encapsulated into **SitesModule**, and we would like to use **/sites/add**, **/sites/edit**, and **/sites/delete** instead of the original links.

This is when child routing comes into the picture. The sample in the **Exercise-09-02** folder contains the files that set up the new routing for **SitesModule**.

In Chapter 3, we had a host component—**AppComponent**—for displaying the site management views. In the new sample **AppComponent** is the host for the app-level routes, so we need a new component within **SitesModule** for this role. Let's create a new component, **SitesComponent**, as shown in Listing 9-8.

Listing 9-8: sites/site.component.ts (Exercise-09-02)

```
import {Component} from '@angular/core';

@Component({
  template: `
    <router-outlet></router-outlet>
  `
})
export class SitesComponent {
}
```

Because the mere purpose of **SitesComponent** is to provide the host view for other site management related components, we do not even need to declare a selector. Though we can give **SitesComponent** a selector, we will never use it in markup, since we only *navigate* to **SitesComponent**.

Our aim is to set up a scenario that uses the URLs shown in Figure 9-7.

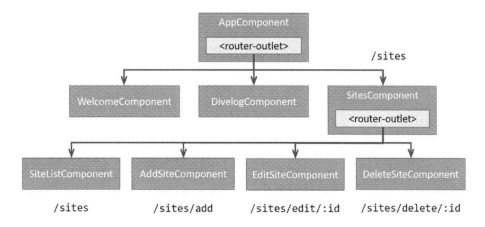

Figure 9-7: Hierachical routing

First, we need to declare the routes used within **SitesComponent**. We create a new file, **sites.routing.ts**, within the **sites** folder, as shown in Listing 9-9.

Listing 9-9: sites.routing.ts (Exercise-09-02)

```
import {Routes, RouterModule} from '@angular/router';
import {SitesComponent} from './sites.component';
import {SiteListComponent} from './site-list.component';
import {AddSiteComponent} from './add-site.component';
import {EditSiteComponent} from './edit-site.component';
import {DeleteSiteComponent} from './delete-site.component';

export const sitesRoutes: Routes = [
  {
    path: 'sites',
    component: SitesComponent,
    children: [
      { path: '', component: SiteListComponent },
      { path: 'add', component: AddSiteComponent },
      { path: 'edit/:id', component: EditSiteComponent },
      { path: 'delete/:id', component: DeleteSiteComponent }
    ]
  }
];

export const sitesRoutingModule = RouterModule.forChild(sitesRoutes);
```

Here, we specify that **SitesComponent** manages the "sites" path, and the route has child paths, as declared in the **children** property. We utilize the **RouterModule.forChild()** method to create the routing module for sites.

Remember, at the app level we invoked the **RouterModule.forRoot()** method. Here, we call **RouterModule.forChild()** only to register additional child routes. The Component Router combines the application level routes with the site management specific routes. Thus, later we can define new feature-specific routes without modifying the app-level route configuration.

Second, we import **sitesRoutingModule** into **SitesModule** (Listing 9-10) so that the Component Router can access all objects to provide routing within the Sites feature set.

Listing 9-10: sites/sites.module.ts (Exercise-09-02)

```
import {NgModule} from '@angular/core';
import {BrowserModule} from '@angular/platform-browser';
import {RouterModule} from '@angular/router';

import {SitesComponent} from './sites.component';
import {SiteListComponent} from './site-list.component';
import {AddSiteComponent} from './add-site.component';
import {EditSiteComponent} from './edit-site.component';
import {DeleteSiteComponent} from './delete-site.component';
```

```
import {SiteManagementService} from './site-management.service'

import {sitesRoutingModule} from './sites.routes';

@NgModule({
  imports: [
    BrowserModule,
    RouterModule,
    sitesRoutingModule
  ],
  declarations: [
    SitesComponent,
    SiteListComponent,
    AddSiteComponent,
    EditSiteComponent,
    DeleteSiteComponent
  ],
  providers: [SiteManagementService]
})
export class SitesModule { }
```

Third, we update **app.routes.ts** to the new scenario (Listing 9-11).

Listing 9-11: app.routes.ts (Exercise-09-02)

```
import {Routes, RouterModule} from '@angular/router';
import {WelcomeComponent} from './welcome/welcome.component';
import {DiveLogComponent} from './logs/dive-log.component';
import {SiteListComponent} from './sites/site-list.component';

import {sitesRoutes} from './sites/sites.routes';

const routes: Routes = [
  { path: 'divelogs', component: DiveLogComponent },
  { path: '', pathMatch: 'full', component: WelcomeComponent },
  ...sitesRoutes
];

export const routingModule = RouterModule.forRoot(routes);
```

In the previous sample we used this routing array:

```
const routes: Routes = [
  { path: 'divelogs', component: DiveLogComponent },
  { path: 'sites', component: SiteListComponent },
  { path: '', pathMatch: 'full', component: WelcomeComponent }
];
```

In Listing 9-11, we replaced the "sites" path we the content of **sites/sites.routes.ts**. We removed the original "sites" entry, and with the help of the "**...**" TypeScript operator, we appended the new "sites" path with its children.

The "..." operator flattens the elements of the array as if we had written this:

```
const routes: Routes = [
  { path: 'divelogs', component: DiveLogComponent },
  { path: '', pathMatch: 'full', component: WelcomeComponent },
  {
    path: 'sites',
    component: SitesComponent,
    children: [
      { path: '', component: SiteListComponent },
      { path: 'add', component: AddSiteComponent },
      { path: 'edit/:id', component: EditSiteComponent },
      { path: 'delete/:id', component: DeleteSiteComponent }
    ]
  }
];
```

Now, the app handles the routing as we expect. When your run it, you can try that all these paths work properly: **/sites**, **/sites/add**, **/sites/edit/1**, **/sites/delete/2**. You can type them into the browser's address bar, and the router navigates to the corresponding components that display their view.

Nonetheless, within the Sites feature area, you cannot use the navigation links associated with the buttons. Because they are not modified according to the new routing scenario, they will not work.

Updating the Links

Why do not the old links work? In **Exercise-03-13**, we used the links highlighted here:

```
<div class="row">
  <div class="col-sm-12">
    <a class="btn btn-primary btn-lg"
      routerLink="/add">
      Add new site
    </a>
  </div>
</div>
<h2>List of Dive Sites</h2>
<div class="row" ...>
  <!-- ... -->
  <div class="col-sm-4" style="margin-top: 5px;">
    <div class="pull-right">
      <a class="btn btn-warning btn-sm"
        [routerLink]="['/edit', site.id]">
        Edit
      </a>
```

```
      <a class="btn btn-danger btn-sm"
        [routerLink]="['/delete', site.id]">
        Delete
      </a>
    </div>
  </div>
</div>
```

The slash character in **/add**, **/edit**, and **/delete** redirects the router to invalid links. Instead of **/add**, **/edit**, and **/delete**, they should go to **/sites/add**, **/sites/edit** and **/sites/delete**, respectively.

We could change **/add** to **/sites/add**, and so on, but setting a link to an absolute path would not be a good idea. Later, if we moved the dive site management feature into another feature area or replaced the "sites" route, we would have to update every associated link one-by-one.

Fortunately, we can use relative paths. Thus, just removing the slash characters would solve this issue, as Listing 9-12 shows.

Listing 9-12: site-list.template.html (Exercise-09-03)

```
<div class="container">
  <h1>List of Dive Sites</h1>
  <div class="row">
    <div class="col-sm-12"
      [style.margin-bottom.px]="12">
      <a class="btn btn-primary btn-lg"
        routerLink="add">
        Add new site
      </a>
    </div>
  </div>
  <div class="row" *ngFor="let site of sites; let e=even; let f=first; let
l=last"
    ywActionable="#aaaaaa" (onAction)="edit(site.id)"
    [ngClass]="{ evenRow: e }"
    [class.topRow]="f"
    [class.bottomRow]="l">
    <div class="col-sm-8">
      <h4>{{site.id}}: {{site.name}}</h4>
    </div>
    <div class="col-sm-4" style="margin-top: 5px;">
      <div class="pull-right">
        <a class="btn btn-warning btn-sm"
          [routerLink]="['edit', site.id]">
          Edit
        </a>
```

```
          <a class="btn btn-danger btn-sm"
            [routerLink]="['delete', site.id]">
            Delete
          </a>
        </div>
      </div>
    </div>
  </div>
```

With this tiny modification, we can access the links from the dive sites list. Nonetheless, we have to modify the links in the other views, too.

In **Exercise-03-13**, the Cancel button of any views redirected the user back to the **/list** link, to the list view. Now, we do not have the **/list** path anymore; the list view is right at the root of the Sites feature area. To go back from **/sites/add** to **/sites**, we can use the "**..**" relative link, as shown in Listing 9-13.

Listing 9-13: add-site.template.html (Exercise-09-03)

```
<div class="container">
  <h3>Specify the name of the dive site:</h3>
  <div class="row">
    <div class="col-sm-6">
      <input #siteNameBox class="form-control input-lg" type="text"
        placeholder="site name"
        (keyup)="siteName=siteNameBox.value"
        (keyup.enter)="added()" />
    </div>
  </div>
  <div class="row" style="margin-top: 12px;">
    <div class="col-sm-6">
      <button class="btn btn-success btn"
        (click)="add()"
        [disabled]="!siteName">
        Add
      </button>
      <a class="btn btn-warning btn"
        routerLink="..">
        Cancel
      </a>
    </div>
  </div>
</div>
```

The Edit and Delete views are a bit different from Add. When we step back from a link such as **/sites/edit/2** to **/sites**, it is not enough to apply the "**..**" relative path, we have to use "**../..**", as shown in Listing 9-14.

Listing 9-14: edit-site.template.html (Exercise-09-03)

```
<div class="container">
  <h3>Edit the name of the dive site:</h3>
  <div class="row">
    <div class="col-sm-6">
      <input #siteNameBox class="form-control input-lg" type="text"
        [value]="siteName"
        placeholder="site name"
        (keyup)="siteName=siteNameBox.value"
        (keyup.enter)="save()" />
    </div>
  </div>
  <div class="row" style="margin-top: 12px;">
    <div class="col-sm-6">
      <button class="btn btn-success btn"
        (click)="save()"
        [disabled]="!siteName">
        Save
      </button>
      <a class="btn btn-warning btn"
        routerLink="../..">
        Cancel
      </a>
    </div>
  </div>
</div>
```

The Add, Edit, and Delete views support saving information. After the successful action, the app goes back to the List view. In **Exercise-03-13** we used the **navigate()** method of the router to change the route programmatically:

```
...
export class AddSiteComponent {
  siteName: string;

  constructor(
    private siteService: SiteManagementService,
    private router: Router) { }

  add() {
    this.siteService.addSite({id: 0, name:this.siteName});
    this.router.navigate(['/list']);
  }
}
```

By default, **navigate()** uses an absolute path. However, besides the first argument that represents the new destination link, it accepts a second argument—an object with the type of **NavigationExtras**—, which can be used—among other purposes—to set a relative path.

Listing 9-15 shows how we can use **navigate()** in **AddSiteComponent** to return to the list view.

Listing 9-15: edit-site.template.html (Exercise-09-03)

```
import {Component, EventEmitter} from '@angular/core';
import {Router, ActivatedRoute} from '@angular/router';

import {SiteManagementService} from './site-management.service'

@Component({
  selector: 'add-site-view',
  templateUrl: 'app/sites/add-site.template.html'
})
export class AddSiteComponent {
  siteName: string;

  constructor(
    private siteService: SiteManagementService,
    private route: ActivatedRoute,
    private router: Router) { }

  add() {
    this.siteService.addSite({id: 0, name:this.siteName});
    this.router.navigate(['..'], {relativeTo: this.route});
  }
}
```

We can inject an **ActivatedRoute** instance into the class constructor, which contains information about a component loaded in an outlet. We utilize this instance in the **add()** method to navigate to the "**..**" route relative to the current one.

> *NOTE: The **NavigationExtras** and **ActivatedRoute** interfaces provide useful features. Visit their API documentation pages for more details.*

In the Edit and Delete views—just like in the corresponding component templates—we use the "**../..**" path:

```
...
this.router.navigate(['../..'], {relativeTo: this.route});
...
```

> *NOTE: You can find the completed app in the **Exercise-09-03** folder.*

A Short Recap of Route Parameters

The Edit and Delete views utilize route parameters that identify the dive site entry to deal with, such as **/sites/edit/1**, **/sites/delete/2**. In the route definitions, we declared these parameters with a colon prefix:

```
...
children: [
  { path: '', component: SiteListComponent },
  { path: 'add', component: AddSiteComponent },
  { path: 'edit/:id', component: EditSiteComponent },
  { path: 'delete/:id', component: DeleteSiteComponent }
]
...
```

As shown in Listing 9-16, we can access this parameter through the **params** object of the route's **snapshot** property.

Listing 9-16: edit-site.component.ts (Exercise-09-03)

```
...
export class EditSiteComponent {
  siteId: number;
  siteName: string;
  private parSub: any;

  constructor(
    private siteService: SiteManagementService,
    private route: ActivatedRoute,
    private router: Router
  ) {
    this.siteId = this.route.snapshot.params['id'];
    this.siteName = this.siteService
      .getSiteById(this.siteId).name;
  }
  // ...
}
```

Using Guards

By default, there is no restriction on how users can navigate within the application. This might be an issue if we need to prevent a user from navigating to a certain link because of the lack of her permissions. Similarly, if we were just filling in a form, we would be annoyed if we could accidentally navigate away without saving the data or confirming the action.

The Component Router has a concept of *guards* to manage these situations. We can create our guards that control the router's behavior.

As of this writing, the router supports three guards.

#1. The **CanActivate** guard tells the router whether the user can *navigate to a route*. We can use this guard to check if a user is authorized. Moreover, we can redirect her to the login page.

#2: The `CanDeactivate` guard tells the router whether the user can *navigate away from the current route*. We can use this moment to save pending changes or ask the user to confirm discarding the changes.

#3. The third guard, **Resolve**, tells the router that we want to retrieve some data before navigating to a route. Prefetching the data could be useful to avoid displaying forms with empty content while they are being loaded.

In this section, we will examine how we can improve the sample app with the `CanActivate` and `CanDeactivate` guards.

User Authorization

Each user has its personal dive log. Thus we change the application so that only authorized users can access the Dive Logs menu. If an anonymous user navigates to this function, we want to redirect him or her to the Login page (Figure 9-8).

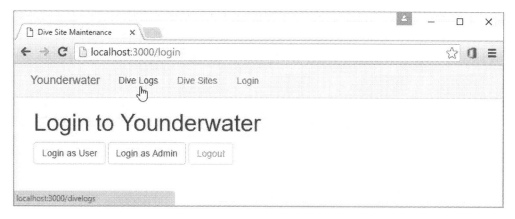

Figure 9-8: The login page

Let's implement this scenario with the help of the `CanActivate` guard. First, we create a `LoginComponent` that handles the login request. For the sake of simplicity, we create a fake login feature that does not require a username and a password: as Figure 9-8 shows, we emulate logging in a simple user or an administrator.

> **NOTE**: *You can find the modified app in the* **Exercise-09-04** *folder. I put all authorization related sources into the* **app/Login** *folder.*

The template of the login page is simple (Listing 9-17).

Listing 9-17: login.template.html (Exercise-09-04)

```
<div class="container">
  <h1>Login to Younderwater</h1>
  <button class="btn btn-default"
    (click)="login('User')"
    [disabled]="authService.loggedInUser">
    Login as User
  </button>
  <button class="btn btn-default"
    (click)="login('Admin')"
    [disabled]="authService.loggedInUser">
    Login as Admin
  </button>
  <button class="btn btn-default"
    (click)="logout()"
    [disabled]="!authService.loggedInUser">
    Logout
  </button>
  <div class="row" [style.margin-top.px]="12">
    <div class="col-sm-12">
      <img [hidden]="!inProgress" src="images/progressring.gif" />
    </div>
  </div>
</div>
```

We emulate that the login takes time, about one second, so the template displays a progress ring while the logic checks the user's identity. **LoginComponent** uses the **UserAuthService** object to carry out the authorization. The component has two operations, **login()** and **logout()**, respectively (Listing 9-18).

Listing 9-18: login.component.ts (Exercise-09-04)

```
import {Component} from '@angular/core';
import {Router} from '@angular/router';
import {UserAuthService} from './user-auth.service';

@Component({
  selector: 'login',
  templateUrl: 'app/login/login.template.html'
})
export class LoginComponent {
  inProgress = false;
  constructor(public authService: UserAuthService,
    public router: Router) {
  }

  login(userName: string) {
    this.inProgress = true;
    this.authService.login(userName).then(() => {
      let redirectUrl = this.authService.redirectUrl
```

```
          ? this.authService.redirectUrl
          : '/';
      this.router.navigate([redirectUrl]);
      this.inProgress = false;
    })
  }

  logout() {
    this.inProgress = true;
    this.authService.logout().then(() => {
      this.inProgress = false;
    })
  }
}
```

The **login()** method accepts a user name, and passes this name to the **UserAuthService** instance. Observe, we use promises to handle the asynchronous nature of the authorization.

After the successful login we redirect the user to the page set in the **redirectUrl** of the authentication service, provided such a URL is specified; otherwise, we send the user to the Welcome page.

Listing 9-19: user-auth.service.ts (Exercise-09-04)

```
import {Injectable} from '@angular/core';

@Injectable()
export class UserAuthService{
  loggedInUser: string = null;
  redirectUrl: string;

  login(userName: string) {
    return new Promise((resolve, reject) => {
      setTimeout(() => {
        this.loggedInUser = userName;
        resolve();
      }, 1000);
    });
  }

  logout() {
    return new Promise((resolve, reject) => {
      setTimeout(() => {
        this.loggedInUser = null;
        this.redirectUrl = null;
        resolve();
      }, 200);
    });
  }
}
```

Due to its fake behavior, the code of **UserAuthService** is pretty short (Listing 9-19).

Creating a CanActivate Guard

The **CanActivate** guard is represented by the interface that has the same name. This interface has a single method, **canActivate()**, which returns a boolean value. True indicates that the user can navigate to the route; false means that the router prevents the user from visiting the route.

To create a guard, we have to declare an injectable class that implements **CanActivate**, as shown in Listing 9-20.

Listing 9-20: logged-in.guard.ts (Exercise-09-04)

```
import {Injectable} from '@angular/core';
import {CanActivate, Router}
  from '@angular/router';
import {ActivatedRouteSnapshot, RouterStateSnapshot }
  from '@angular/router';
import {UserAuthService} from './login/user-auth.service';

@Injectable()
export class LoggedInGuard implements CanActivate {
  constructor(
    private authService: UserAuthService,
    private router: Router) {
  }

  canActivate(route: ActivatedRouteSnapshot,
    routerState: RouterStateSnapshot) {
      if (this.authService.loggedInUser) {
        return true;
      }
      this.authService.redirectUrl = routerState.url;
      this.router.navigate(['/login']);
      return false;
  }
}
```

The constructor of **LoggedInGuard** accepts a **UserAuthService** and a **Router** instance. The **canActivate()** method receives a snapshot of the current route and the state of the router. If we have a user already logged in, the guard returns **true**. Otherwise, it redirects the user to the Login page and retrieves **false** to prevent the router to access the link.

Because the guard is invoked before the navigation is carried out, **routerState.url** is always set to the route we are navigating from. Outside of **UserAuthService**, we set the **redirectUrl** property so that we can send back users to the URL from which they are requesting the authorization.

> **NOTE:** *Observe, I put the* `Logged-in.guard.ts` *file directly into the* **app** *folder, as we use the guard's functionality there. Nonetheless, you could put it into the* **app/Login** *folder, as well.*

We intend to use the login functionality from the application root module, so we encapsulate the login-related objects into a separate module, **LoginModule** (Listing 9-21).

Listing 9-21: login.module.ts (Exercise-09-04)

```
import {NgModule, ModuleWithProviders} from '@angular/core';
import {BrowserModule} from '@angular/platform-browser';
import {LoginComponent} from './login.component';
import {UserAuthService} from './user-auth.service';

@NgModule({
  imports: [BrowserModule],
  declarations: [LoginComponent],
  providers: [UserAuthService]
})
export class LoginModule {
}
```

To apply the **CanActivate** guard, we need to configure the routing (Listing 9-22).

Listing 9-22: app.routes.ts (Exercise-09-04)

```
import {Routes, RouterModule} from '@angular/router';
import {WelcomeComponent} from './welcome/welcome.component';
import {DiveLogComponent} from './logs/dive-log.component';
import {SiteListComponent} from './sites/site-list.component';

import {sitesRoutes} from './sites/sites.routes';
import {LoggedInGuard} from './logged-in.guard';
import {UserAuthService} from './login/user-auth.service';
import {LoginComponent} from './login/login.component';

const routes: Routes = [
  { path: 'divelogs', component: DiveLogComponent,
    canActivate: [LoggedInGuard] },
  { path: 'login', component: LoginComponent },
  { path: '', pathMatch: 'full', component: WelcomeComponent },
  ...sitesRoutes
];

export const routingProviders = [
  LoggedInGuard,
  UserAuthService
];
export const routingModule = RouterModule.forRoot(routes);
```

We intend to guard the Dive Logs menu, so we assign **LoggedInGuard** to the "divelogs" route with the **canActivate** property.

As you can observe, **canActivate** accepts an array, since you can pass multiple guards. Should any of them disallow navigation, you would not be able to reach the route.

To allow to visit the login page, we add the "login" path to the route definitions.

While we use the app, we need to use both **LoggedInGuard** and **UserAuthService**. Angular can inject them into the requesting components only when we create providers for them. To keep the service providers used exclusively for routing separated, we put them into the **routingProviders** array and export the array. To complete the authorization story, we update the application root module (Listing 9-23), import **LoginModule**, and declare the providers we need for routing.

Listing 9-23: app.modules.ts (Exercise-09-04)

```
import {NgModule} from '@angular/core';
import {BrowserModule} from '@angular/platform-browser';
import {AppComponent} from './app.component';

import {WelcomeModule} from './welcome/welcome.module';
import {DiveLogModule} from './logs/dive-log.module';
import {SitesModule} from './sites/sites.module';
import {LoginModule} from './login/login.module';

import {routingModule, routingProviders} from './app.routes';

@NgModule({
  imports: [
    BrowserModule,
    WelcomeModule,
    DiveLogModule,
    SitesModule,
    LoginModule,
    routingModule
  ],
  declarations: [
    AppComponent
  ],
  providers: [routingProviders],
  bootstrap: [AppComponent]
})
export class AppModule { }
```

Start the app, and try to navigate to the **/divelogs** path. Check the Dive log menu and type the path to the address bar, too. You can experience that the app will redirect you to the login page. Once you log in either as a user or an admin, you find yourself in the Dive Logs page.

Using Authorization in the Sites Page

Now, that we have user authorization, we can utilize it to check user rights in other actions, too. The **Exercise-09-05** folder contains a sample that allows using the Delete button of the Dive Sites list only for administrative users (Figure 5-9).

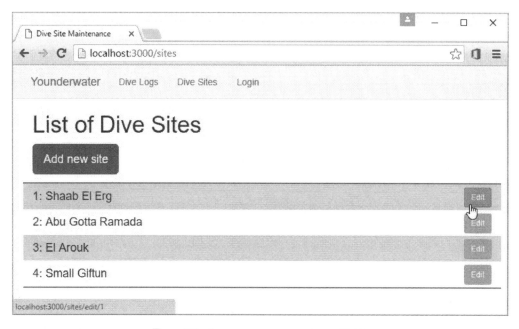

Figure 5-9: Anonymous users cannot access Delete

With a subtle modification in the **site-list.template.html** file (Listing 9-24), and in **SiteListComponent** (Listing 9-25), we can benefit from the functionality of **UserAthService**.

Listing 9-24: site-list.template.html (Exercise-09-05)

```
...
<div class="col-sm-4" style="margin-top: 5px;">
  <div class="pull-right">
    <a class="btn btn-warning btn-sm"
      [routerLink]="['edit', site.id]">
      Edit
    </a>
    <a class="btn btn-danger btn-sm"
      *ngIf="authService.loggedInUser == 'Admin'"
      [routerLink]="['delete', site.id]">
      Delete
    </a>
  </div>
</div>
...
```

Listing 9-25: site-list.template.html (Exercise-09-05)

```
...
export class SiteListComponent {
  sites: DiveSite[];

  constructor(
    public authService: UserAuthService,
    private siteService: SiteManagementService,
    private router: Router
  ) {
    this.sites = siteService.getAllSites();
  }
}
...
```

Creating a CanDeactivate Guard

While we are adding a new site to the list, or editing the name of an existing one, we might navigate away without saving. Adding a **CanDeactivate** guard to the affected routes, we can ask the user whether he or she wants to discard the changes.

The Component Router provides a **CanDeactivate<>** generic interface to vote whether a particular component allows navigating away. This interface defines a single method, **canDeactivate()** that returns a boolean flag.

The generic nature of **CanDeactivate<>** is very useful. We can provide a guard that implements the general way of route deactivation, while components can add their specific code to customize the generic behavior. Listing 9-26 shows how the sample in the **Exercise-09-06** folder implements such a confirmation logic.

Listing 9-26: deactivable.guard.ts (Exercise-09-06)

```
import { Injectable }    from '@angular/core';
import { CanDeactivate } from '@angular/router';

export interface DeactivableComponent {
 allowLeave: () => boolean;
 confirmText: () => string;
}

@Injectable()
export class CanDeactivateGuard implements
CanDeactivate<DeactivableComponent> {
  canDeactivate(component: DeactivableComponent) : boolean |
Promise<boolean> {
    if (component.allowLeave()) return true;
```

```
    var text = component.confirmText() ||
      'Do you want to discard these changes?'
    return new Promise<boolean>(resolve => {
      return resolve(window.confirm(text));
    });
  }
}
```

Here, The **DeactivableComponent** interface defines the contract a component should support so that we can involve it into the deactivation logic. The **allowLeave()** function retrieves a boolean, the **true** value of which indicates that we can navigate away. The **confirmText()** method retrieves a confirmation message.

The **CanDeactivateGuard** class implements **CanDeactivate<>** with the **DeactivableComponent** used as the type parameter. According to this setup, the **canDeactivate()** method accepts an argument that supports the **DeactivableComponent** interface so that we can implement the confirmation dialog logic right within the guard method.

Whenever the component needs confirmation, we pop up the appropriate confirmation message—or a default one, if the component does not provide any—, and continue with the answer provided.

Observe the **boolean | Promise<boolean>** return type of **canDeactivable()**. This construct is a TypeScript feature, called *union type*. It means that this method can retrieve either a boolean value or a promise that resolves to a boolean. Thus, we can write **canDeactivate()** to support both synchronous and asynchronous operations.

To benefit from **CanDeactivateGuard**, we have to modify **AddSiteComponent** and **EditSiteComponent** so that they implement **DeactivableComponent**. We need to write only a few lines of code, as Listing 9-27 and Listing 9-28 show.

Listing 9-27: add-site.component.ts (Exercise-09-06)

```
...
import {DeactivableComponent} from '../deactivable.guard';
// ...
export class AddSiteComponent implements DeactivableComponent {
  siteName = '';
  origSiteName = '';

  constructor(
    private siteService: SiteManagementService,
    private route: ActivatedRoute,
    private router: Router) { }

  cancel() {
    this.router.navigate(['..'], {relativeTo: this.route});
  }
```

```
  add() {
    this.siteService.addSite({id: 0, name:this.siteName});
    this.origSiteName = this.siteName;
    this.router.navigate(['..'], {relativeTo: this.route});
  }

  allowLeave = () =>this.siteName == this.origSiteName;
  confirmText = () => null;
}
```

Listing 9-28: edit-site.component.ts (Exercise-09-06)

```
...
import {DeactivableComponent} from '../deactivable.guard';
// ...
export class EditSiteComponent implements DeactivableComponent {
  siteId: number;
  siteName: string;
  private parSub: any;
  private origSiteName: string;

  constructor(
    private siteService: SiteManagementService,
    private route: ActivatedRoute,
    private router: Router
  ) {
    this.siteId = this.route.snapshot.params['id'];
    this.siteName = this.siteService
      .getSiteById(this.siteId).name;
    this.origSiteName = this.siteName;
  }

  save() {
    this.siteService.saveSite({id: this.siteId, name:this.siteName});
    this.origSiteName = this.siteName;
    this.router.navigate(['../..'], {relativeTo: this.route});
  }

  allowLeave = () =>this.siteName == this.origSiteName;
  confirmText = () => "Do you want to discard the modifications?";
}
```

To let **DeactivableGuard** intercept component navigation, we need to configure the routing accordingly, as shown in Listing 9-29.

Listing 9-29: sites.routes.ts (Exercise-09-06)

```
...
export const sitesRoutes: Routes = [
  {
    path: 'sites',
    component: SitesComponent,
    children: [
      { path: '', component: SiteListComponent },
      { path: 'add', component: AddSiteComponent,
        canDeactivate: [CanDeactivateGuard] },
      { path: 'edit/:id', component: EditSiteComponent,
        canDeactivate: [CanDeactivateGuard] },
      { path: 'delete/:id', component: DeleteSiteComponent }
    ]
  }
];

export const sitesRoutingModule = RouterModule.forChild(sitesRoutes);
```

The `canDeactivate` property of the route definition accepts an array of guards. If any of these guards disallow navigating away from the route, the router will stay at the current URL.

To allow Angular inject `CanDeactivateGuard`, we need to change the `routingProviders` array (Listing 9-30).

Listing 9-30: app.routes.ts (Exercise-09-06)

```
...
import {CanDeactivateGuard} from './deactivable.guard';
// ...
export const routingProviders = [
  LoggedInGuard,
  UserAuthService,
  CanDeactivateGuard
];
...
```

Start the app, go to the Dive Sites menu and edit a site. Change the site name and click Cancel. As Figure 9-10 shows, the `CanDeactivateGuard` displays the confirmation dialog.

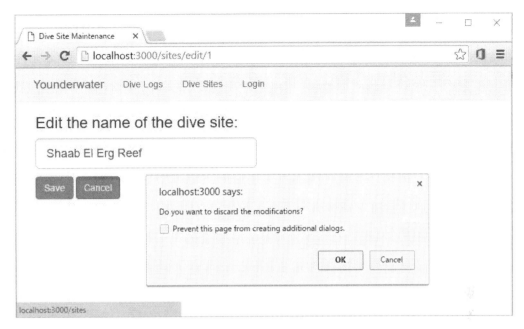

Figure 9-10: The CanDeactivateGuard in action

A Few More Remarks On Guards

Though earlier I mentioned that guards return a boolean value, they can return a
`Promise<boolean>` or an `Observable<boolean>` value, too. The framework knows that when it
receives a `Promise<boolean>` or an `Observable<boolean>`, it should wait for the result of the
asynchronous execution.

There is a third type of guard, `Resolve`, which can be used to retrieve data before moving to the
new route. The `Resolve` interface declares the `resolve()` method that returns the data you expect
before the navigation.

Asynchronous Component Loading

A large application may have dozens or hundreds of feature and shared modules. Most of them
are only rarely accessed and utilized. Why load them? It just takes time and consumes memory—
without even using the module.

The Component Router helps you in such scenarios: instead of loading all modules at bootstrap
time, you can postpone the loading of modules by the time they are first used.

The `Exercise-09-07` folder contains a modified version of the previous app. Instead of loading
the Sites feature set at the startup of the app, the Component Router will load the module the
first time you navigate to any route that would require a component in the Sites module.

Let's see what changes we need to support the on-demand loading scenario.

We have to make two slight modifications in the **SitesModule** component. First, as Listing 9-31 shows, we should change **BrowserModule** to **CommonModule**. Using **BrowserModule** raises an error message. Second, because the child routes will not be loaded statically into the application root module, they should be loaded in **sites.module.ts**

Listing 9-31: sites.module.ts (Exercise-09-07)

```
import {NgModule} from '@angular/core';
import {CommonModule} from '@angular/common';
import {RouterModule} from '@angular/router';

import {SitesComponent} from './sites.component';
import {SiteListComponent} from './site-list.component';
import {AddSiteComponent} from './add-site.component';
import {EditSiteComponent} from './edit-site.component';
import {DeleteSiteComponent} from './delete-site.component';
import {SiteManagementService} from './site-management.service'

import {sitesRoutingModule} from './sites.routes';

@NgModule({
  imports: [
    CommonModule,
    RouterModule,
    sitesRoutingModule
  ],
  declarations: [
    SitesComponent,
    SiteListComponent,
    AddSiteComponent,
    EditSiteComponent,
    DeleteSiteComponent
  ],
  providers: [SiteManagementService]
})
export class SitesModule { }
```

As of this writing, I have not got an explanation from the Angular team yet why we need to do that, and whether this is a final feature or just an intermediate state.

In the previous sample, we added the child routes of Sites to the application routing statically:

```
import {sitesRoutes} from './sites/sites.routes';
//...
const routes: Routes = [
  { path: 'divelogs', component: DiveLogComponent, canActivate:
[LoggedInGuard] },
  { path: 'login', component: LoginComponent },
  { path: '', pathMatch: 'full', component: WelcomeComponent },
  ...sitesRoutes
```

```
];
```

Instead of doing this, we declare the "sites" path with the **loadChildren** configuration property, as shown in Listing 9-32.

Listing 9-32: app.routes.ts (Exercise-09-07)

```
...
const routes: Routes = [
  { path: 'divelogs', component: DiveLogComponent,
    canActivate: [LoggedInGuard] },
  { path: 'login', component: LoginComponent },
  { path: 'sites',
    loadChildren: 'app/sites/sites.module#SitesModule' },
  { path: '', pathMatch: 'full', component: WelcomeComponent },
];

export const routingProviders = [
  LoggedInGuard,
  UserAuthService,
  CanDeactivateGuard
];
export const routingModule = RouterModule.forRoot(routes);
```

Take a look at the value of the **loadChildren** property. You can recognize that it names the module file that hosts the Sites module. After the path to the file, the # character separates the class name of the **SitesModule** NgModule. This route definition instructs the router that it should load the **app/sites/sites.module** file and activate **SitesModule**.

> *NOTE: A single file may contain more than one NgModules. The notation used in **LoadChildren** makes it possible to identify the single NgModule instance to activate.*

So far, the **sites.routes.ts** file defined a single root with the "sites" path, and it was statically added to **app.routes.ts**. However, now we have a "sites" path definition in **app.routes.ts**, so we need to modify the path of the route definition to an empty string in **sites.routes.ts** to use proper child paths (Listing 9-33).

Listing 9-33: sites.routes.ts (Exercise-09-07)

```
export const sitesRoutes: Routes = [
  {
    // --- Originally it was path: 'sites'
    path: '',
    component: SitesComponent,
    children: [
      { path: '', component: SiteListComponent },
      { path: 'add', component: AddSiteComponent,
        canDeactivate: [CanDeactivateGuard] },
```

```
    { path: 'edit/:id', component: EditSiteComponent,
      canDeactivate: [CanDeactivateGuard] },
    { path: 'delete/:id', component: DeleteSiteComponent }
    ]
  }
];
```

Without this modification the router would add an extra "sites" path, thus, for example, it would navigate to **/sites/sites/add** instead of **/sites/add**.

We are almost ready, but still need to take care of one important detail. We have to remove **SitesModule** from **AppModule**. Without this action, the framework would load **SitesModule** at bootstrap time (Listing 9-34), and our whole preparation for asynchronous loading would be in vain.

Listing 9-34: app.module.ts (Exercise-09-07)

```
import {NgModule} from '@angular/core';
import {BrowserModule} from '@angular/platform-browser';
import {AppComponent} from './app.component';

import {WelcomeModule} from './welcome/welcome.module';
import {DiveLogModule} from './logs/dive-log.module';
// !!! import {SitesModule} from './sites/sites.module';
import {LoginModule} from './login/login.module';

import {routingModule, routingProviders} from './app.routes';

@NgModule({
  imports: [
    BrowserModule,
    WelcomeModule,
    DiveLogModule,
    // !!! SitesModule,
    LoginModule,
    routingModule
  ],
  declarations: [
    AppComponent
  ],
  providers: [routingProviders],
  bootstrap: [AppComponent]
})
export class AppModule { }
```

Start the app. When the browser displays the page, it shows the Welcome panel. In your browser's development tools, you can check that page does not load source files from within the **app/sites** folder (Figure 9-11).

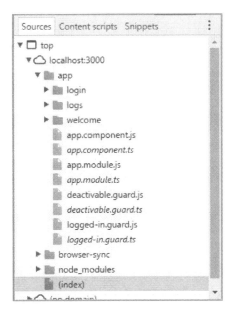

Figure 9-11: The source files for Sites are not yet loaded

Now, click the Dive Sites menu. Due to the routing configuration, the Component Router observes that the **SitesModule** NgModule is not loaded yet, and it loads the module, then navigates to **SitesComponent**. As Figure 9-12 shows, now the page includes the files in the **app/sites** folder.

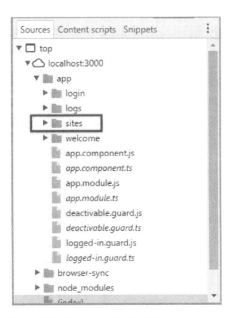

Figure 9-12: The source files for Sites are now loaded

Summary

The Angular Component Router understands URLs we enter in the address bar and navigates to a component associated with that URL. To component can access the parameters and options passed in the address, process them and render the view accordingly.

Angular supports hierarchical routing; a routed component can be the host of other routed child components.

With guards, you can control whether users can navigate to a particular route, or navigate away from that. With the asynchronous loading feature of the Component Router, you can load NgModules on-demand.

Chapter 10: Pipes

WHAT YOU WILL LEARN IN THIS CHAPTER

Getting to know what Angular pipes are

Using the built-in pipes

Creating your own custom pipes

Understanding how you can use pure and impure pipes

Angular has a concept, *pipes*, which can be used to transform data as it is passed from the model represented by a component or a directive to a view. This concept plays an important role to separate the model of the app from the view that displays it. A pipe provides flexible data transformations—so you can show the data in a variety of formats depending on the view.

Let's assume the model provides a data value, namely **2.34**. If this data is a currency value—let's say the unit price of a product—, it should be displayed accordingly. If you are in the US, it is displayed as **$2.34**. Should you visit the same web page from Hungary, you see **2,34 Ft**. Whose responsibility is to transform the value of **2.34** to a currency format depending on the locale of the user? Putting it into the model would add extra unnecessary complexity.

In most of your apps, you rather have several or several dozen occurrences of such conversion than a single one.

You can separate data transformations into reusable pipes, and utilize this transformation logic in the templates of your components. This separation not only increases the flexibility of the app but also makes it more readable and maintainable.

In this chapter, you will learn about the predefined Angular pipes, and get acquainted creating your owns, too.

Using the Predefined Angular Pipes

Angular defines several useful pipes that you can use out of the box. Table 10-1 summarizes them.

Table 10-1: Predefined Angular Pipes

Pipe	Description
async	Subscribes to an **Observable** or **Promise** and returns the latest value it has emitted
currency	Formats a number as local currency
date	Formats a date value to a string based on the requested format
json	Transforms any input value using **JSON.stringify()**
lowercase	Transforms text to lowercase
number	Formats a number as local text. It uses locale-specific configurations that are based on the active locale.
percent	Formats a number as local percent
slice	This pipe creates a new **List** or **String** containing only a subset (slice) of the input elements
uppercase	Transforms text to uppercase

Let's see these pipes in action! To demonstrate their behavior, we are going to use the dive log example that you already used in several chapters of this book. For the sake of brevity, in listings, I will show only extracts from the source files that focus on using pipes. Of course, you will be able to try entire samples, which can be found in the source code download folder of this chapter.

> **NOTE**: *Using the* **async** *pipe requires some more knowledge on Observables. You will find more details in* Chapter 11, The Async Pipe

Formatting Strings

It is easy to use pipes, as the code extract in Listing 10-1 shows. This sample adds the **uppercase** and **lowercase** pipes to expressions that display a dive log entry.

Listing 10-1: dive-log.template.html (Exercise-10-01)

```
...
<div class="col-sm-4"
  *ngFor="let dive of dives">
  <h3>{{dive.site | uppercase}}</h3>
```

```
    <h4>{{dive.location | lowercase}}</h4>
    <h2>{{dive.depth}} feet | {{dive.time}} min</h2>
</div>
...
```

Pipes can be used within Angular template expressions. Here, inside the interpolation expressions, we flow the `dive.site` and `dive.location` values through the `uppercase` and `lowercase` pipes, respectively. Pipes are separated by the *pipe operator*—a vertical bar character ("|")—from the expression they are about to transform. There's nothing surprising; dive log entries are displayed as expected (Figure 10-1).

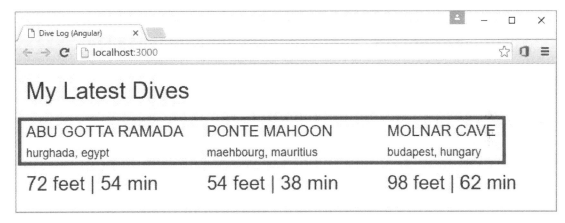

Figure 10-1: Using the uppercase and lowercase pipes

It's important to be aware of that only those "|" characters are parts of a pipe that are within an Angular expression. In the following markup snippet, the vertical bar is a simple character to display—and it does nothing to do with pipes:

```
<h2>{{dive.depth}} feet | {{dive.time}} min</h2>
```

Formatting Numbers

Let's assume you want to display the average time spent under water, and the average bottom depth together with dive log entries. The markup showing this would look something like this:

```
...
<div class="row">
  <div class="col-sm-12">
    <h3>
      Average dive time:
      {{avgTime()}}
      minute
    </h3>
    <h3>
      Average depth:
```

```
      {{avgDepth()}}
      feet
    </h3>
  </div>
</div>
...
```

It's perfect for its goal, but you will not like the output as shown in Figure 10-2—there are too many unnecessary digits after the decimal point.

Average dive time: 51.333333333333336 minute

Average depth: 74.66666666666667 feet

Figure 10-2: There are too many digits after the decimal point

With the help of the **number** pipe, you can easily get rid of the unnecessary digits, as shown in Listing 10-2.

Listing 10-2: dive-info.template.html (Exercise-10-02)

```
<div class="row">
  <div class="col-sm-12">
    <h3>
      Average dive time:
      {{avgTime() | number: '3.2-2'}}
      minute
    </h3>
    <h3>
      Average depth:
      {{avgDepth() | number}}
      feet
    </h3>
  </div>
</div>
```

The **number** pipe accepts an optional string parameter (as typed after the colon) that specifies the format of the number. This string has the following format:

```
{minInt}.{minFract}-{maxFract}
```

In this expression, *{minInt}* represents the minimum number of integer digits to use. The *{minFract}* and *{maxFract}* values are for the minimum and maximum number of digits after fraction. If we omit the format string, it defaults to "1.0-3". Figure 10-3 shows the output generated by Listing 10-2.

Average dive time: 051.33 minute

Average depth: 74.667 feet

Figure 10-3: Using the number pipe

Generating JSON Output

You need it rarely—maybe for technical sites or for debugging purposes—, but you can use the **json** pipe to transform an object to its JSON string representation, as shown in Listing 10-3.

Listing 10-3: dive-log.template.html (Exercise-10-03)

```
...
<div class="row">
  <div class="col-sm-4"
    *ngFor="let dive of dives">
    <pre>{{dive | json}}</pre>
  </div>
</div>
...
```

Though the result (Figure 10-4) is not quite user-friendly—unless you are the developer who needs exactly this output—, but it demonstrates the effect of utilizing the **json** pipe.

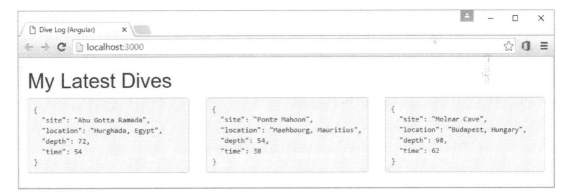

Figure 10-4: Using the json pipe

Formatting Dates

If there is a type that benefits from formatting, **Date** is definitely that one. Not surprisingly, there is an Angular pipe for this purpose, **date**, which accepts an optional string argument, the format of the date. The parameter value can be a format string assembled from date/time components (for example, "MM/dd/yyyy, G"), or a predefined format (for example, "longDate"). Listing 10-4 shows an example.

Listing 10-4: dive-log.template.html (Exercise-10-04)

```
...
<div class="col-sm-4"
  *ngFor="let dive of dives; let e=even; let o=odd;">
    <h3>{{dive.site}}</h3>
    <h4>{{dive.location}}</h4>
    <h2>{{dive.depth}} feet | {{dive.time}} min</h2>
    <h4 *ngIf="e">
      {{dive.divedOn | date: 'short'}}
    </h4>
    <h4 *ngIf="o">
      {{dive.divedOn | date: 'MMMM dd, yyyy'}}
    </h4>
</div>
...
```

The markup demonstrates the "short" predefined format, and an exact format ("MMMM dd, yyyy"). Figure 10-5 shows the result of applying the **date** pipe.

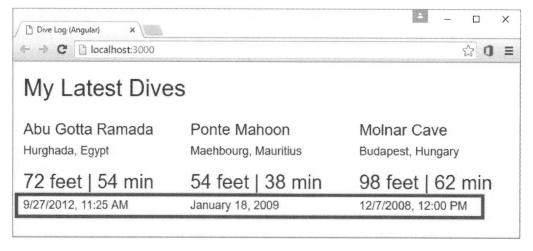

Figure 10-5: Using the date pipe

HINT: *You can get more details about the date format string from the official Angular API reference* (https://angular.io/docs/ts/latest/api/).

Formatting Currency Values

We often need to display currency values. Not surprisingly, the **currency** pipe is the best way to do that, as the sample code in Listing 10-5 demonstrates.

Listing 10-5: dive-log.template.html (Exercise-10-05)

```
...
<div class="col-sm-4"
  *ngFor="let dive of dives; let e=even; let o=odd;">
    <h3>{{dive.site}}</h3>
    <h4>{{dive.location}}</h4>
    <h2>{{dive.depth}} feet | {{dive.time}} min</h2>
    <h4>
      Extra paid:
      <span *ngIf="e">
        {{dive.extraPaid | currency:'EUR':'true'}}
      </span>
      <span *ngIf="o">
        {{dive.extraPaid | currency: 'USD':false:'1.2-2'}}
      </span>
    </h4>
</div>
...
```

The pipe has three optional parameters. The first is the currency code—according to ISO 4217—such as "USD", "EUR", "HUF", etc. The second parameter is a Boolean flag that indicates whether to use the currency symbol (true) or the currency code (false). The third parameter uses the same format as the number pipe to specify the format of the numeric value of the currency.

Figure 10-6 shows the result of the currency formats used in the listing.

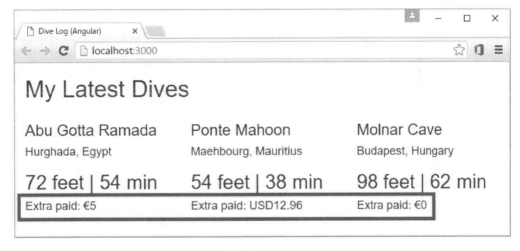

Figure 10-6: Using the currency pipe

Displaying Percentage Values

You can display percentage values with the **percent** pipe, as shown in Figure 10-7. The depth values are presented in percentages, proportionally to the average depth of dives.

Figure 10-7: The percent pipe in action

As Listing 10-6 demonstrates, the **percent** pipe has an optional parameter following the same formatting rule as the argument of the **number** pipe.

Listing 10-6: dive-log.template.html (Exercise-10-06)

```
...
<div class="col-sm-4"
  *ngFor="let dive of dives">
  <h3>{{dive.site}}</h3>
  <h4>{{dive.location}}</h4>
  <h2>{{dive.depth / avgDepth() | percent:'1.2-2'}}  | {{dive.time}}
min</h2>
</div>
...
```

Displaying a Subset of Lists

The **slice** pipe provides an easy way to limit a list to a subset. As its name suggests, it utilizes the **slice()** JavaScript function to create a subset of the input.

The pipe accepts two parameters, **start**, and **end**; both of them are optional. If you omit both of them, the full list is retrieved. These parameters use the same semantics as the arguments of the **slice()** function—they accept both positive and negative integers.

Listing 10-7 shows the template of **NumberListComponent**—see its source code in the **Exercise-10-07** folder—that applies the **slice** pipe with its arguments bound to **<input>** tags.

Listing 10-7: number-list.template.html (Exercise-10-07)

```html
<div class="container-fluid">
  <h1>My Numbers</h1>
  <div class="row" style="margin-bottom: 12px;">
    <div class="col-sm-2">
      <input #startBox class="form-control input-lg"
        placeholder="Start"
        value="0"
        (keyup)="start=toValue(startBox.value, 0)" />
    </div>
    <div class="col-sm-2">
      <input #endBox class="form-control input-lg"
        placeholder="End"
        value="100"
        (keyup)="end=toValue(endBox.value, null)" />
    </div>
  </div>
  <span *ngFor="let num of numbers | slice:start:end"
    class="number">
    {{num}}
  </span>
</div>
```

Figure 10-8 and Figure 10-9 show the **slice** pipe in action.

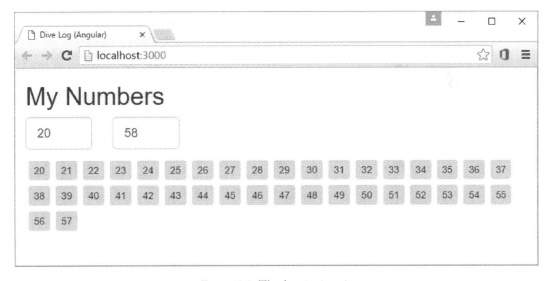

Figure 10-8: The slice pipe in action

Figure 10-9: The slice pipe with negative arguments

The **slice** pipe works not only with lists but with strings, too. The **Exercise-10-08** folder contains an additional sample that demonstrates using **slice** with a string (Figure 10-10).

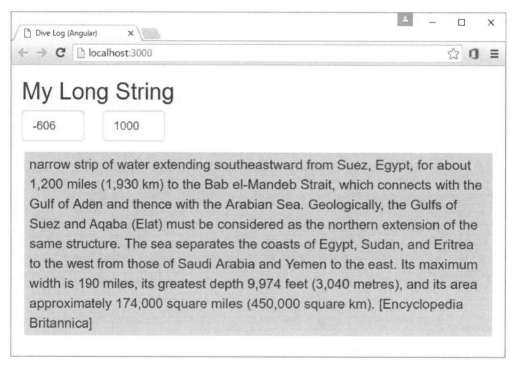

Figure 10-10: The slice pipe applied to a string

Combining Pipes and Expressions

Pipes and expressions can be combined. When you apply a pipe to an expression, the result will be another expression—so it seems that you can take the vertical bar into account as an operator within an expression. Unfortunately, the Angular documentation does not mention if the vertical bar is treated as an operator during template expression parsing, and what its precedence is. Though, the documentation explicitly says that pipes can be chained.

Nonetheless, we can combine them. Use parentheses to be explicit when creating an expression that contains multiple pipes. Listing 10-8 shows a sample.

Listing 10-8: dive-info.template.html (Exercise-10-09)

```
<div class="col-sm-12">
  <h3>
    Average dive time:
    {{ (avgTime() | number: '3.2-2') + ' minutes' | uppercase }}
  </h3>
  <h3>
    Average depth:
    {{ avgDepth() | number | slice:1:4 }}
    feet
  </h3>
</div>
</div>
```

The first highlighted snippet concatenates two expressions: a number rounded to two fractional digits, and an uppercase string. The second highlighted expression chains two pipes. First, the **number** pipe is applied and then **slice**. The **number** pipe results in a string, so **slice** creates a substring from the second character (start index is 1) to the fourth. After **number**, the average depth value is represented as "74.667". Adding **slice** to the chain results in "4.6".

The result of applying these expressions is shown in Figure 10-11.

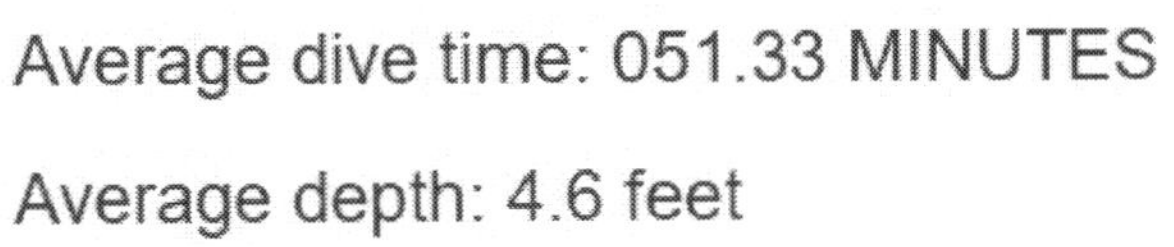

Average dive time: 051.33 MINUTES

Average depth: 4.6 feet

Figure 10-11: The result of pipes and expressions combined

Creating Custom Pipes

Back in *Chapter 1, Filtering Dive Log Entries with a Pipe*, you already learned that you could easily create custom pipes. In this section, you will understand how to add your custom pipes to apps.

As you already got acquainted with components, directives, and services, pipes are classes decorated with particular metadata.

Formatting a Dive Log Entry Item

First, let's see a sample pipe that formats a dive log entry. Listing 10-09 defines a filter that simplifies the formatting of the depth and time information of a dive log entry with a custom pipe. As you have already seen in several code listings, the depth and bottom time data of a dive log entry is formatted this way:

```
{{depth}} feet | {{time}} min
```

The listing creates a pipe that carries out this formatting. Moreover, it allows specifying the separator string between the depth and bottom time values. So, with the **diveData** pipe, we can simplify the expression above to this one:

```
{{ dive | diveData }}
```

Let's see the code that defines the **diveData** pipe (Listing 10-09)!

Listing 10-09: dive-data.pipe.ts (Exercise-10-10)

```
import {Pipe, PipeTransform} from '@angular/core';
import {DiveLogEntry} from './dive-log-entry';

@Pipe({name: 'diveData'})
export class DiveDataPipe implements PipeTransform {
  transform(dive: DiveLogEntry, separator: any = ' | '): string {
    if (dive && dive.depth && dive.time) {
      return `${dive.depth} feet${separator}${dive.time} min`;
    }
    return '(no dive data)';
  }
}
```

The **@Pipe()** decorator assigns metadata to the **DiveDataPipe** class. It signs that this class acts as a pipe, and it should be associated with the "diveData" name in Angular expressions. The **DiveDataPipe** class implements the **PipeTransform** interface, which defines the **transform()** method. When Angular applies a pipe, it passes the value to the pipe (the value to the left to "|") as the first parameter of **transform()**, and then the pipe arguments declared in the pipe expression as subsequent parameters.

> *TypeScript: The language supports default function parameters. The definition of the **separator** parameter tells the compiler that if no parameter value is passed to **separator**, the ' | ' string should be used.*

The body of the **transform()** function is very straightforward.

To utilize the pipe in the component template, we need to add it either to the **pipes** metadata property of the **DiveLogComponent**, or to the declarations of **AppModule**. In this sample, I modified the **app.module.ts** (Listing 10-10).

Listing 10-10: dive-log.component.ts (Exercise-10-10)

```
import {NgModule} from '@angular/core';
import {BrowserModule} from '@angular/platform-browser';

import {DiveLogComponent} from './dive-log.component';
import {DiveDataPipe} from './dive-data.pipe';

@NgModule({
  imports: [BrowserModule],
  declarations: [
    DiveLogComponent,
    DiveDataPipe
  ],
  bootstrap: [DiveLogComponent]
})
export class AppModule { }
```

If we did not do this, Angular would raise an error about the missing pipe. Fortunately, we do not need to add the pipes property with the classes of predefined pipes; the bootstrap mechanism does this for us.

The **diveData** pipe is ready to try, of course, we need to apply it in **DiveLogComponent**'s template:

```
...
<div class="col-sm-4"
  *ngFor="let dive of dives">
  <h3>{{dive.site}}</h3>
  <h4>{{dive.location}}</h4>
  <h2>{{dive | diveData: ' :: '}}</h2>
</div>
...
```

As Figure 10-12 shows, the **diveData** pipe works as we expect.

The concept of pipes makes it possible to create very flexible transformations. It does not constrain the input data types or the pipe arguments you can use. However, you should be very careful with pipes. If you use them carelessly, your app may have a severe performance penalty.

Nothing prohibits you from using slow and resource consuming operations in a filter. Theoretically—and even practically—you may talk to a backend service within a pipe, and even wait for the response. It is very easy to add a pipe to the markup—you need to type only a few characters—without thinking about the performance consequences.

Figure 10-12: The diveData pipe in action

In the next section, you will learn important details about pipes in connection to the Angular change detection mechanism. It will help you understand crucial design considerations about pipes.

Pipes and Change Detection

You already learned about the Angular change detection mechanism. You know that the framework looks for data-bound value changes. The process that is responsible for observing them runs after every JavaScript event, including keyboard events—every keystroke generates such an event—mouse moves, the arrival of responses from the backend server, timer ticks, and so on. You can imagine that this may be very expensive by means of resources consumed while this mechanism runs.

Angular tries to keep this cost as low as possible by applying faster—and thus cheaper—change detection algorithm with pipes. If you do not know how it works, you may be surprised.

Missing the Change in Input Values

Earlier you worked with the dive log entry app that contained a search box to filter the list of dives according to search criteria. The **Exercise-10-11** folder contains a modified sample that helps demonstrate the peculiarities of change detection with pipes. Run this app and go through these steps to observe an issue:

1. As the app starts, it displays three log items. Click the Add extra dives button—and you see a log with six entries (Figure 10-13).

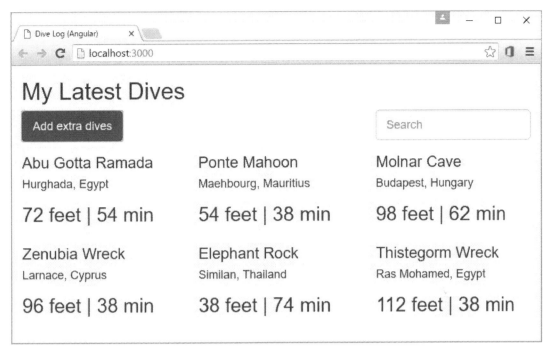

Figure 10-13: Dive log with six entries

2. Type "mo" in the search box and the list will be filtered to two items that contain "mo" somewhere in their details (Figure 10-14).

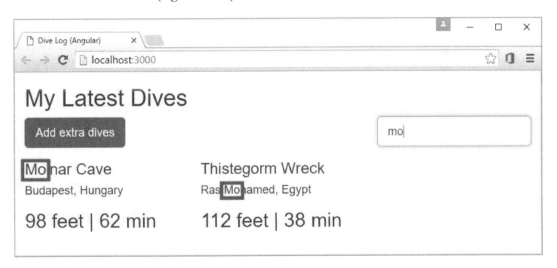

Figure 10-14: Filtered dive log entries

3. Refresh the browser page. Now, first type "mo" in the search box—you see only a single item displayed—, and then click the Add extra dives button. You should see the same two items as in Figure 10-14, but amazingly, there is still only a single one shown (Figure 10-15).

Figure 10-15: Only one dive log entry is displayed instead of two.

4. Empty the search box. The page displays all the six items you saw earlier in Figure 10-13.

Why did not show up the app the second dive log entry matching with "mo" in the latest step? Is it an Angular bug? No. To understand what is going on, you should know the behavior of the change detection mechanism.

The code adds new dive log entries to the **dives** component property with the **addDives()** method:

```
addDives() {
  DiveLogEntry.ExtraDives.forEach(item => {
    this.dives.push(item);
  });
}
```

When using pipes, Angular checks two things to determine whether there is any change. These are the input of the pipe and its arguments. In this sample, the arguments did not change when you clicked the Add extra dives button. The input array changed, for the button click added new entries.

However, when checking the change of input, by default, Angular tests whether the *reference to the input* has changed. In this example, the input of the pipe is the **dives** array:

```
<div class="col-sm-4"
  *ngFor="let dive of dives | contentFilter:searchBox.value">
  <!-- ... -->
</div>
```

Although the **addDives()** method mutated the **dives** array, it did not change the reference to **dives**, so the change detection process did not observe change after the invocation of the method.

Fixing the Change Detection Issue

To let the framework detect the change, we replace the content of the **dives** array instead of mutating it. Such an operation explicitly modifies the reference to **dives**, and thus Angular will observe it. The sample in the **Exercise-10-12** folder adds an extra line—as highlighted—to **addDives()**:

```
addDives() {
  this.dives = this.dives.slice(0);
  DiveLogEntry.ExtraDives.forEach(item => {
    this.dives.push(item);
  });
}
```

Now, when you repeat the scenario above, the app will work as you expect. It will detect the newly added dive log entries, and will display all items that match the search criteria.

Pure and Impure Pipes

Angular pipes can be divided into two categories: they can be either *pure* or *impure*. These differ only in the way the framework tests whether their inputs are changed.

Pure pipes are executed only when their input is changed to a primitive JavaScript value (string, number, Boolean) or an object reference. Angular does not check any changes within objects, thus replacing an element of an array, mutating a property of an object, adding or removing list items or properties does not trigger pipe execution.

Impure pipes are executed when any mutation in the input is detected. Not only object references are compared but all object properties in full depth.

By default, all pipes are pure, so Angular uses the cheap change detection mechanism, and this is why the first sample (**Exercise-10-11**) produced the unexpected behavior.

We can set the **pure** metadata property of the **@Pipe** annotation to declare the pureness of the pipe explicitly. The **Exercise-10-13** folder contains a clone of **Exercise-10-11** and fixes it by declaring the **ContentFilterPipe** as impure (Listing 10-11).

Listing 10-11: content-filter.pipe.ts (Exercise-10-13)

```
import {Pipe, PipeTransform} from '@angular/core';
import {DiveLogEntry} from './dive-log-entry';

@Pipe({name: 'contentFilter', pure: false})
export class ContentFilterPipe implements PipeTransform {
  transform(value: DiveLogEntry[], searchFor: string) : DiveLogEntry[] {
    // ...
  }
}
```

Setting the **pure** metadata property to **false** turns the pipe into impure.

Replacing Input Object versus Impure Pipes

We implemented two alternative ways to let Angular detect the changes of the **dives** input array used in tandem with the **contentFilter** pipe. The first one replaced the input array instead of mutating it; the second applied an impure pipe. There comes a natural question: which solution is preferred?

There is no definite answer to this question. Evidently, implementing impure pipes is the easiest way: you only need to set the **pure** metadata property to **false**, and that is all. When the input object is not very large and complex, this is a fair solution.

Nonetheless, you should know that change detection of impure pipes may be very expensive. If for example, you have an array, Angular compares the previous state of the array with the new state—I mean the state of the array when the next change detection process runs—, which might be the same as the previous one.

If the array is not replaced and its length is the same as it was previously, Angular traverses through all items looking for any changed item. It compares element properties one-by-one. If those are composite objects, the comparison goes on diving deeply—and recursively—into those objects.

The result of this check might be much ado about nothing: the framework spends a lot of resources the guess out that there are no changes at all.

So, I suggest you replace the input of the pipe when you suspect that applying impure pipes would result in wasting resources.

> *HINT: In many cases, not using pipes at all could be the best way to avoid potential performance issues. Instead of creating a pipe, you create a function that transforms the input the same manner your pipe would do and apply this function to the input.*

Pure and Impure Behavior

Pure pipes should always use pure functions—functions that do not cause side effects, and always return the same output, provided they are fed with the same input.

Side effects may raise error messages regarding expressions that have changed after they were checked. The **Exercise-10-15** folder contains a sample that demonstrates this plight. In this code, the filter is bound to the component's **search** property—unlike in the previous samples when the filter was directly bound to the value of the **searchBox <input>** tag. In its body, the modified **ContentPipeFilter** sets the search property of the component to an empty string, so the pipe creates a side effect that relates back to itself (Listing 10-12).

Listing 10-12: content-filter.pipe.ts (Exercise-10-14)

```
import {Pipe, PipeTransform} from '@angular/core';
import {DiveLogEntry} from './dive-log-entry';
import {DiveLogComponent} from './dive-log.component';


@Pipe({name: 'contentFilter'})
export class ContentFilterPipe implements PipeTransform {

  transform(value: DiveLogEntry[], logComp: DiveLogComponent, searchFor:
string) : DiveLogEntry[] {
    logComp.search = '';
    if (!searchFor) return value;
    searchFor = searchFor.toLowerCase();
    return value.filter(dive =>
      dive.site.toLowerCase().indexOf(searchFor) >= 0 ||
      dive.location.toLowerCase().indexOf(searchFor) >= 0 ||
      dive.depth.toString().indexOf(searchFor) >= 0 ||
      dive.time.toString().indexOf(searchFor) >= 0);
  }
}
```

When you run this sample and apply a search that changes the list of dive log entries, you get an "Expression has changed after it was checked" error message (Figure 10-16)—caused by the side effect.

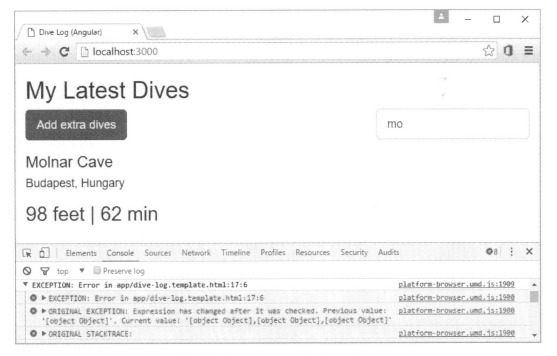

Figure 10-16: Error due to side effects in the pipe

> **NOTE**: *The method that I passed a component reference to a pipe—a reference that uses the very same pipe—is an anti-pattern, it may be the source of serious issues, so avoid it. Here I used it only for the sake of demonstrating the side effect caused by the pipe.*

Summary

Angular pipes can be used to transform data as it is passed from a component to the view. The framework has a few predefined filters that can be used to convert single values as well as collections.

It's incredibly easy to create your custom pipes—you have no limitations what kind of data transformation you can do with them.

Angular change detection tries to be frugal with the system resources, so it detects changes in pipe input by comparing references. With impure pipes, you can ask the framework for comparing objects deeply and observe property mutations.

In the next chapter, you will learn nitty-gritty details about the asynchronous observable pattern.

Chapter 11: Observables

WHAT YOU WILL LEARN IN THIS CHAPTER

Understanding the asynchronous observable pattern

Getting acquainted with the fundamental traits of observables

Using observables for inter-component communication

Understanding how the `Http` service leverages observables

Applying the `async` pipe

Earlier in the book, in *Chapter 4, Using the Http Service*, you learned that Angular leverages the *asynchronous observables* pattern; it utilizes this technique in the `Http` service. You also used this method in *Chapter 7, Communication Through a Service*, when you created a service object to use it as an intermediary between a parent and a child component.

In this chapter, you will understand the basic concepts behind this pattern, and learn how to use observables with Angular.

Understanding Observables

Developers and frameworks (especially UI libraries) have been using the *observer design pattern* for a long time. The essence of this pattern is that an object, called the *subject*, maintains a list of dependent objects, called *observers*. When the state of the subject changes, it notifies all of its observers—usually calling a method of an observer object.

Due to the nature of method calling, when the subject changes its state, it must invoke the notification methods of its observers synchronously and sequentially, and it has to wait while those methods return. For example, the subject can be that an `<input>` element receives a `keyup` event. Observers are those objects that attach their event handler methods to `<input>`.

> *HINT: Search the web for "observer design pattern" and you find dozens of pages that explain you the concept with all nitty-gritty details.* http://reactivex.io/ *is a great starting point.*

In the era of increased user experience expectations, we do not tolerate when the synchronous nature of apps punishes us with lagging or temporarily frozen UIs. We expect that a subject can notify observers asynchronously so that the slow answer of a particular observer could not hinder other observers from doing their jobs and thus could not damage the user experience.

Reactive Programming and Data Streams

As the time goes on, the events of subject changes produce a data stream (Figure 11-1). For example, as an `<input>` element receives keystrokes, the sequence of **keyup** events constitutes a data stream of event parameters. (Each little, rounded box in the figure represents a **keyup** event.)

time

Figure 11-1: Data stream

There is a relatively new style of programming, *reactive programming*, which leverages asynchronous data streams. In reactive programming, a data stream is a sequence of ongoing events that are ordered in time. Events can represent three things: a *value*, an *error*, or a *signal of completion*, respectively. For example, when we use an `<input>` element to type in an integer number, **keyup** events may create a value when we press a digit key, error, when an invalid character is used, and a signal of completion for the Enter key (Figure 11-2).

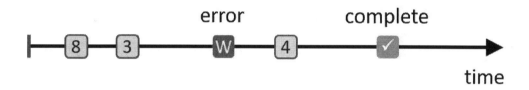

time

Figure 11-2: Data stream in reactive programming

In reactive programming, we never catch these events synchronously—there is no way to do this. We can capture the events only asynchronously through functions. To process values, errors, and completion, we apply separate functions. In many cases, we do not care about errors and completion; we only process the values the stream emits.

To catch the events, we need to subscribe to a stream. With the subscription, we define the handler functions (observers) to be notified.

A Data Stream Sample

To demonstrate this approach in practice, let's examine a short sample. The **Exercise-11-01** folder contains an app in which we collect mouse coordinates as the user moves the mouse over a virtual mousepad on the screen (Figure 11-3).

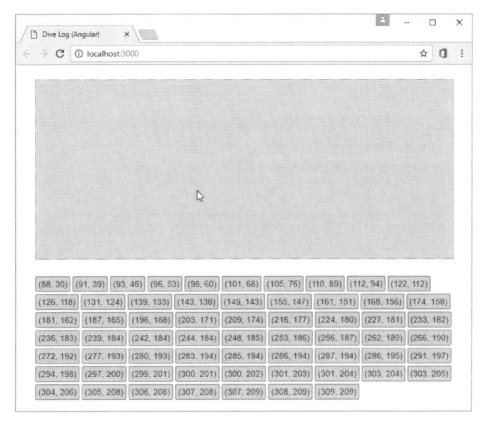

Figure 11-3: A data stream sample in action

The fundamental piece of this sample is the highlighted part of the **AppComponent** class (Listing 11-1).

Listing 11-1: app.component.ts (Exercise-11-01)

```
import {Component, AfterViewInit} from '@angular/core';
import {Observable} from 'rxjs/Observable';
import 'rxjs/add/observable/fromEvent';

@Component({
  selector: 'yw-app',
  template: `
    <div class="container">
      <div class="col-sm-12 mousepad"
        id="mousepad">
      </div>
      <log-board [messages]="messages"></log-board>
    </div>
  `,
  // ...
})
```

```
export class AppComponent implements AfterViewInit {
  messages: any[] = [];

  log(message: any) {
    this.messages.push(message)
  }

  ngAfterViewInit() {
    let mousepad = document.getElementById('mousepad');
    let mouseMove$ = Observable
      .fromEvent(mousepad, 'mousemove');

    mouseMove$.subscribe(
      (m: MouseEvent) => {
        this.log(`(${m.clientX}, ${m.clientY})`);
      }
    );
  }
}
```

At the beginning of the file, we import a few definitions from the RxJS library. This entire library contains tons of objects, methods, and operators, but we intend to use only a few of them. So, instead of importing everything, we only load the definition of **Observables** and the **fromEvent()** function. **Observable** itself represents the data stream. The **Observable.fromEvent()** operation creates a data stream that returns event objects coming from **mousemove** event of the DOM element represented by **mousepad**. By convention, we add a **$** suffix to variable names that stand for a data stream; this is why we use **mouseMove$**.

The **subscribe()** method of an observable data stream accepts the three observer function, one for values, errors, and the completion signal, in this very order. In the app, we pass only the values handler that logs mouse coordinates.

Observables have useful traits that are important to understanding what is going on when we utilize them: they are lazy, cancellable, retriable, and composable. Let's see these traits one by one!

Observables Are Lazy

The **Exercise-11-02** folder helps to demonstrate why we say that observables are lazy. Listing 11-2 shows a modified version of the previous sample. Here, we use the **do()** operator to declare what we intend to carry out when we observe a **MouseEvent**.

Listing 11-2: app.component.ts (Exercise-11-02)

```
import {Component, AfterViewInit} from '@angular/core';
import {Observable} from 'rxjs/Observable';
import 'rxjs/add/observable/fromEvent';
import 'rxjs/add/operator/do';
```

```
@Component({
  // ...
})
export class AppComponent implements AfterViewInit {
  messages: any[] = [];

  log(message: any) {
    this.messages.push(message)
  }

  ngAfterViewInit() {
    let mousepad = document.getElementById('mousepad');
    let mouseMove$ = Observable
      .fromEvent(mousepad, 'mousemove')
      .do((m: MouseEvent) => {
        this.log(`(${m.clientX}, ${m.clientY})`);
      });

    mouseMove$.subscribe();
  }
}
```

This program works just like the previous sample. However, here, instead of passing a values observer function, we display the values within the action of the **do()** operator.

When we run the app, we can check that it does the same as **Exercise-11-01**. Seemingly, the **mouseMove$.subscribe()** does not do any useful thing because we do not pass any observer function to the method call.

Let's comment out the **mouseMove$.subscribe()** call. Running the application now does not produce any output as if the code caught no mouse events. This behavior is a consequence of the laziness of observables. The following statement just prepares an observable object, but does not ignite it:

```
let mouseMove$ = Observable
  .fromEvent(mousepad, 'mousemove')
  .do((m: MouseEvent) => {
    this.log(`(${m.clientX}, ${m.clientY})`);
  });
```

Here, **mouseMove$** is just a declaration of our intention. This observable definition does not start watching for events and processing them—unless we subscribe to it with **mouseMove$.subscribe()**.

Independent Observers

Let's invoke **subscribe()** twice:

```
...
mouseMove$.subscribe();
mouseMove$.subscribe();
...
```

The running app now has two subscriptions for the same observable definition. Both of them are attached to the same source—to the stream of **mousemove** events—, but each of them has its dedicated observer that logs the mouse coordinates. As Figure 11-4 shows, each observer processes the events independently—and that is why every move is logged twice.

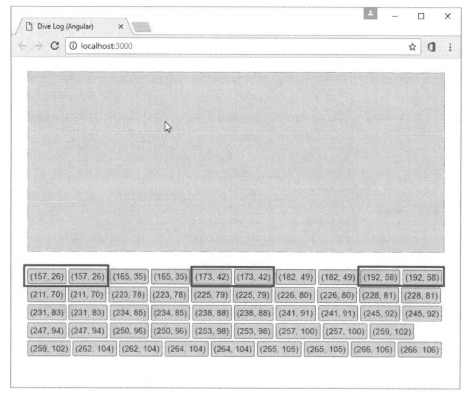

Figure 11-4: Event are logged by both observers

Observables Are Cancellable and Retriable

We can anytime unsubscribe from an observable, and then re-subscribe to it. Unsubscribing means that we cancel observing the data stream, re-subscribing means that we return to receiving events from the stream again.

The sample in the **Exercise-11-03** folder extends the previous app. When we click the mouse button, it toggles the subscription state of the stream. The first click cancels, the second re-subscribes, and so on. Listing 11-3 highlights the key changes.

Listing 11-3: app.component.ts (Exercise-11-03)

```
import {Component, OnInit} from '@angular/core';
import {Observable} from 'rxjs/Observable';
import 'rxjs/add/observable/fromEvent';

@Component({
  selector: 'yw-app',
  template: `
    <div class="container">
      <div class="col-sm-12 mousepad"
        id="mousepad"
        [class.subscribed]="subscription"
        (click)="toggleSubscribe()">
      </div>
      <log-board [messages]="messages"></log-board>
    </div>
  `,
  // ...
})
export class AppComponent implements OnInit {
  messages: any[] = [];
  subscription: any = null;
  mouseMove$: Observable<any>;

  log(message: any) {
    this.messages.push(message)
  }

  ngOnInit() {
    let mousepad = document.getElementById('mousepad');
    this.mouseMove$ = Observable
      .fromEvent(mousepad, 'mousemove');

    this.toggleSubscribe();
  }

  toggleSubscribe() {
    if (this.subscription) {
      this.subscription.unsubscribe();
      this.subscription = null;
      this.log('Cancelled');
    } else {
      this.subscription = this.mouseMove$.subscribe(
        (m: MouseEvent) => {
          this.log(`(${m.clientX}, ${m.clientY})`);
        });
```

```
        }
      }
    }
```

The first time `toggleSubscribe()` is called in `ngOnInit()`, it invokes `mouseMove$.subscribe()` as it did in the previous samples. However, this time, it stores the object returned from this call in the `subscription` property. When the first mouse click arrives, `subscription` is used to invoke the `unsubscribe()` method that cancels the subscription. The next mouse click invokes `mouseOver$.subscribe()` again.

Figure 11-5 shows the messages of cancellation.

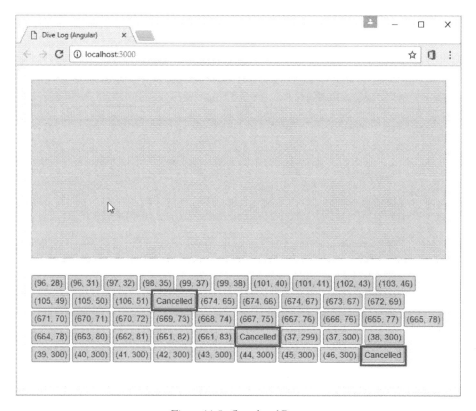

Figure 11-5: Cancel and Retry

As soon as we unsubscribe, the observable stops watching for events and releases resources used while serving the data stream. When we subscribe again, the observable allocates resources again and starts listening to events.

Observables Are Composable

One of the most valuable feature of observables is that we can compose them—we can create an observable from another one. In Listing 11-4, we use the **filter()** operator to transform a data stream into another.

Listing 11-4: app.component.ts (Exercise-11-04)

```
...
import 'rxjs/add/operator/filter';

@Component({
  // ...
})
export class AppComponent implements OnInit {
  messages: any[] = [];
  subscription: any = null;
  mouseOver$: Observable<any>;
  counter = 0;

  log(message: any) {
    this.messages.push(message)
  }

  ngOnInit() {
    let mousepad = document.getElementById('mousepad');
    this.mouseOver$ = Observable
      .fromEvent(mousepad, 'mousemove')
      .filter(value => this.counter++ % 10 == 0);
    this.toggleSubscribe();
  }

  toggleSubscribe() {
    // ...
  }
}
```

The previous examples generated too many outputs, even when we moved the mouse slowly. To reduce the number of events, we reduce the data stream with the **filter()** operator that accepts a predicate—a function that transforms its argument to a boolean value.

In the listing, we use a counter to push through only every tenth event and throwing away the others. Figure 11-6 depicts how the original stream, **fromEvents()**, is transformed by **filter()**.

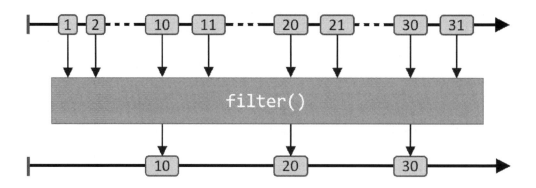

Figure 11-6: Transforming a data stream into another

RxJS defines and implements dozens of operators. You can find a summary of them in this page: http://reactivex.io/documentation/operators.html. When you visit the page and select a particular operator—such as **delay()**, shown in Figure 11-7—, you can get language-specific information. Check the RxJS panel, and follow the links defined there for more details.

Figure 11-7: Getting information about RxJS operators

> **NOTE**: *To add an operator to your source code, import the appropriate source file from the* **rxjs/add/operator** *folder.*

Component Communication with Observables

In the previous sections, we demonstrated samples in the context of a single component. Real apps generally define a data stream in one component and consume it in another. Here, we refactor the mouse event sample to reveal component communication through observables.

Using Subject

This app represents the mousepad with a component, **MousepadComponent**. As Listing 11-5 shows, it utilizes the **Subject<>** generic class to define the stream of events.

Listing 11-5: mousepad.component.ts (Exercise-11-05)

```
import {Component, Input} from '@angular/core';
import {Subject} from 'rxjs/subject';

@Component({
  selector: 'mousepad',
  template: `
    <div class="col-sm-12"
      (mousemove)="onMouseMove($event)"
      (click)="onClick()">
    </div>
  `,
  // ...
})
export class MousepadComponent {
  counter = 0;
  private eventSource = new Subject<MouseEvent>();
  mouseEvents$ = this.eventSource.asObservable();

  onMouseMove(e: MouseEvent) {
    this.counter++;
    if (this.counter % 10 == 0) {
      this.eventSource.next(e);
    }
    if (this.counter % 200 == 0) {
      this.eventSource.error("error");
    }
  }

  onClick() {
    this.eventSource.complete();
  }
}
```

In the introduction, we already learned that the data stream is produced by the events that represent the changes of a *subject*. Accordingly, the **Subject<>** type defines this subject, and its type parameter represents the data that is put into the stream. In the code, with the help of asObservable(), we create an observable, mouseEvents$, from the source of events. We are going to utilize mouseEvent$ to communicate with the other components of the app.

The onMouseMove() method produces a change in the subject's state with invoking the next() method of eventSource, and after receiving 200 mousemove events, it generates an error.

When the user clicks the mousepad, the complete() method raises a signal of completion.

Listing 11-6 shows that AppComponent renders MousepadComponent as a child. We access this child with the @ViewChild(MousepadComponent) decoration through the mousepad property and pass three observer methods to subscribe(), the values, the error, and the completion observers, in this very order.

Listing 11-6: app.component.ts (Exercise-11-05)

```
import {Component, AfterViewInit} from '@angular/core';
import {ViewChild} from '@angular/core';
import {MousepadComponent} from './mousepad.component';

@Component({
  selector: 'yw-app',
  template: `
    <div class="container">
      <mousepad></mousepad>
      <log-board [messages]="messages"></log-board>
    </div>
  `
})
export class AppComponent implements AfterViewInit {
  @ViewChild(MousepadComponent) mousepad: MousepadComponent;
  messages: any[] = [];
  counter = 0;

  ngAfterViewInit() {
    this.mousepad.mouseEvents$.subscribe(
      (m: MouseEvent) => {
        this.log(`(${m.clientX}, ${m.clientY})`);
      },
      () => { this.log("Error!"); },
      () => { this.log("Completed."); }
    )
  }

  log(message: any) {
    this.messages.push(message)
  }
}
```

When you run the app, clicking the mousepad completes the data stream. With no clicks, after 200 moves the stream generates an error. After emitting an error or completion, the stream does not convey new data.

Resuming the Subject

When its **error()** or **complete()** method is called, the subject goes into a stopped state. To resume from this situation, you have to create a new subject instance and subscribe to its data stream again. The **Exercise-11-06** folder contains a modified version of the sample, where we can click a Resume button to retrieve to normal operation after an error or completion. Listing 11-7 and Listing 11-8 highlight the key changes.

Listing 11-7: app.component.ts (Exercise-11-06)

```
...
export class AppComponent implements AfterViewInit {
  @ViewChild(MousepadComponent) mousepad: MousepadComponent;
  messages: any[] = [];
  counter = 0;

  ngAfterViewInit() {
    this.subscribe();
  }

  subscribe() {
    this.mousepad.mouseEvents$.subscribe(
      (m: MouseEvent) => {
        this.log(`(${m.clientX}, ${m.clientY})`);
      },
      () => { this.log("Error!"); },
      () => { this.log("Completed."); }
    )
  }

  log(message: any) {
    this.messages.push(message)
  }

  resume() {
    if (this.mousepad.isStopped) {
      this.mousepad.createStream();
      this.subscribe();
    }
  }
}
```

Listing 11-8: mousepad.component.ts (Exercise-11-06)

```
...
export class MousepadComponent {
  counter: number;
  private eventSource: Subject<MouseEvent>;
  mouseEvents$;

  constructor() {
    this.createStream();
  }

  // ...
  isStopped() {
    return this.eventSource.isStopped;
  }

  createStream() {
    this.counter = 0;
    this.eventSource = new Subject<MouseEvent>();
    this.mouseEvents$ = this.eventSource.asObservable();
  }
}
```

We can test the **isStopped** property of a **Subject<>** instance to check whether the subject is in the stopped state. The **resume()** method of **AppComponent** utilizes this value to decide whether it should recreate the stream. Figure 11-8 shows the sample in action.

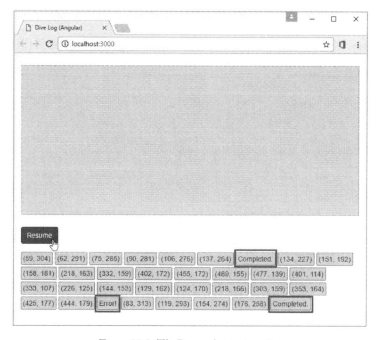

Figure 11-8: The Resume button in action

The Http Service and Observables

In *Chapter 4, Using the Http Service*, we already learned that the **Http** service uses observables. That time, we converted an **Http** call's result into a **Promise** instance so that we can use promise pattern instead of asynchronous observables:

```
getDives() {
  return this.http.get(this.DIVE_LOG_API_URL).toPromise()
    .then(resp => resp.json())
    .catch(err => {
      // ...
      return Promise.reject(errMsg);
    })
};
```

In this section, we break up with promises and see how **Http** works with observables. The **Exercise-11-07** folder contains a modified version of the sample, **Exercise-04-04**, which displayed dive log entries coming from an Azure-hosted website. Listing 11-9 shows the new way **DiveLogApi** class retrieves dive log entries from the backend.

Listing 11-9: dive-log-api.service.ts (Exercise-11-07)

```
import {Injectable} from '@angular/core';
import {Http} from '@angular/http';
import {Observable} from 'rxjs/Observable';
import 'rxjs/add/operator/map';
import 'rxjs/add/operator/catch';
import 'rxjs/add/observable/throw';
import {DiveLogEntry} from './dive-log-entry';

@Injectable()
export class DiveLogApi {
  private DIVE_LOG_API_URL =
    'http://unraveling-ng.azurewebsites.net/api/backendtest/dives';
  constructor (private http: Http) {}

  getDives() {
    return this.http.get(this.DIVE_LOG_API_URL)
      .map(resp => resp.json() || [])
      .catch(err => {
        let errMsg = (err.message)
          ? err.message
          : err.status ? `${err.status}: ${err.statusText}`
                       : 'Server error';
        console.error(errMsg);
        return Observable.throw(errMsg);
      })
  }
}
```

The **getDives()** method invokes **http.get()** to obtain the list of dive log entries through the specified URL. This call—as we learned—returns an observable, namely an **Observable<Response>** instance, where **Response** represents the **Http** response. Just as we did in Chapter 4, we use the **json()** function of **Response** to obtain the result from the body of the reply.

Because we are working with observables, we use the **map()** operator that transforms a stream of **Response** objects into a stream of **DiveLogEntry[]** objects. Should any error happen, the **catch()** operator would convert it into an error with **Observable.throw()**.

At the end of the day, **getDives()** returns a data stream. An object that listens to this stream can observe **DiveLogEntry[]** values, an error, or a signal of completion.

> **NOTE**: *Because the http GET operation we issue returns only a single response, the data stream will produce up to one value.*

Neither **map()** nor **catch()** transforms the signal of completion. The listener that subscribes to the stream coming from **getDives()** can observe the completion of the stream.

Listing 11-10 shows how **DiveLogComponent** consumes the service API with the pattern we used in the previous samples.

Listing 11-10: dive-log.component.ts (Exercise-11-07)

```
import {Component} from '@angular/core';
import {DiveLogEntry} from './dive-log-entry';
import {DiveLogApi} from './dive-log-api.service'

@Component({
  selector: 'divelog',
  templateUrl: 'app/dive-log.template.html'
})
export class DiveLogComponent {
  loading = false;
  dives: DiveLogEntry[];
  errorMessage: string = null;

  constructor(private api: DiveLogApi) {
  }

  refreshDives() {
    this.loading = true;
    this.dives = [];
    this.errorMessage = null;

    this.api.getDives()
      .subscribe(
        (dives: DiveLogEntry[]) => {
          this.dives = dives
```

```
      },
      err => this.errorMessage = err,
      () => this.loading = false
    );
  }
}
```

Using forEach, catch, and then

The **Observables** type provides a fluent interface that we can use instead of **subscribe()**. With the help of **forEach()**, **catch()**, and **then()**, we can transform the consuming site into a more straightforward form, as shown in Listing 11-11.

Listing 11-11: dive-log.component.ts (Exercise-11-08)

```
...
export class DiveLogComponent {
  loading = false;
  dives: DiveLogEntry[];
  errorMessage: string = null;

  constructor(private api: DiveLogApi) {
  }

  refreshDives() {
    this.loading = true;
    this.dives = [];
    this.errorMessage = null;

    this.api.getDives()
      .forEach(
        (dives: DiveLogEntry[]) => {
          this.dives = dives
      })
      .catch(err => this.errorMessage = err)
      .then(() => this.loading = false);
  }
}
```

Instead of passing the values, error, and completion observers to **subscribe()**, we pass them to **forEach()**, **catch()**, and **then()**, respectively.

Retrying Http Requests

When sending an HTTP request to a backend, sometimes we can get an intermittent error. The cause of such an error can be a temporary overload or another short-term issue. In similar scenarios, we can retry to send the request after a short delay. With observables, we can easily resend web requests.

In the previous sample, we accessed a real website to obtain dive log entries. Let's modify the code to generate fake errors so that we can test how observables-based retry works. Listing 11-12 shows the modifications we added to the **DiveLogApi** service.

Listing 11-12: dive-log-api.service.ts (Exercise-11-09)

```
...
@Injectable()
export class DiveLogApi {
  private _count = 0;
  private DIVE_LOG_API_URL =
    'http://unraveling-ng.azurewebsites.net/api/backendtest/dives';

  constructor (private http: Http) {}

  getDives() {
    return this.http.get(this.DIVE_LOG_API_URL)
      .map(resp => {
        if (++this._count % 3 != 0) return resp;
        throw { message: 'Fake error' };
      })
      .map(resp => resp.json() || [])
      .catch(err => {
        let errMsg = (err.message)
          ? err.message
          : err.status ? `${err.status}: ${err.statusText}` : 'Server
error';
        console.error(errMsg);
        return Observable.throw(errMsg);
      })
  }
}
```

Because we cannot modify the backend service, we emulate the error within the **getDives()** API method. The highlighted **map()** operator transforms the original data stream so that it raises a fake error for every third call. The consumers of **DiveLogApi** receive the error and process it as if it came from the backend. As shown in Figure 11-9, **DiveLogComponent** catches the error and displays it on the page.

For resuming the data stream from error, RxJS provides two useful operators, **retry()**, and **retryWhen()**, respectively. These operators simply emit the same values as the source observable, except when they receive an error. The **retry()** method takes an optional counter as an argument and re-subscribes to the original observable, unless the number of unsuccessful retries reaches the specified counter. The **retryWhen()** operator can decide whether it wants to retry the operation at all—for example, depending on the error raised. It returns a modified observable. Whenever this observable emits new data to the stream, it retries the operation with re-subscribing to the source observable.

Figure 11-9: The fake error is displayed

Listing 11-13 demonstrates how `retryWhen()` can be used to resume the backend call after one-second delay.

Listing 11-13: dive-log-api.service.ts (Exercise-11-10)

```
import {Injectable} from '@angular/core';
import {Http, Response} from '@angular/http';
import {Observable} from 'rxjs/Observable';
import 'rxjs/add/operator/map';
import 'rxjs/add/operator/catch';
import 'rxjs/add/observable/throw';
import 'rxjs/add/operator/retrywhen';
import 'rxjs/add/operator/delay';
import {DiveLogEntry} from './dive-log-entry';


@Injectable()
export class DiveLogApi {
  private _count = 0;
  private _isRetry = false;
  private DIVE_LOG_API_URL =
    'http://unraveling-ng.azurewebsites.net/api/backendtest/dives';

  constructor (private http: Http) {}

  getDives() {
    return this.http.get(this.DIVE_LOG_API_URL)
      .map(resp => {
        if (this._isRetry) {
          this._isRetry = false;
          return resp;
        }
        if (++this._count % 3 != 0) return resp;
        throw { message: 'Fake error' };
      })
      .map(resp => resp.json() || [])
```

```
    .retryWhen(source => {
      console.log('Retrying...')
      this._isRetry = true;
      return source.delay(1000)
    })
    .catch(err => {
      let errMsg = (err.message)
        ? err.message
        : err.status ? `${err.status}: ${err.statusText}` : 'Server
error';
      console.error(errMsg);
      return Observable.throw(errMsg);
    })
  }
}
```

The key is the **source.delay(1000)** call that retrieves the **source** observable value after the delay. As soon as this value is emitted, the **retryWhen()** operator subscribes to the original observable again. Note, the class uses the **_isRetry** flag so that in can distinguish retries from the normal calls and handle the counter appropriately.

When you run the **Exercise-11-10** app, you can experience that every third refresh takes about one second longer than the others. This is the delay we built into the logic.

The Async Pipe

Angular implements an unusual impure pipe, **AsyncPipe**—you can use it with the "**async**" name in component templates—, which works with a **Promise** or **Observable** input. This pipe automatically subscribes to the input stream, returns the emitted values, and does the necessary cleanup on the completion of the data stream.

The sample in the **Exercise-11-11** folder uses **AsyncPipe** to display the last **MouseEvent** information. Listing 11-14 shows that the **async** pipe is applied on an **Observable<MouseEvent>** instance that conveys events from **MousepadComponent**.

Listing 11-14: app.component.ts (Exercise-11-11)

```
import {Component, AfterViewInit} from '@angular/core';
import {ViewChild} from '@angular/core';
import {MousepadComponent} from './mousepad.component';

@Component({
  selector: 'yw-app',
  template: `
    <div class="container">
      <mousepad></mousepad>
```

```
      <span class="message">
        {{ getMouseMessage(mousepad.mouseEvents$ | async) }}
      </span>
    </div>
  `,
  styles: [`
    .message {
      display: inline-block;
      background-color: #e0e0e0;
      border: 1px solid #808080;
      padding: 2px 4px;
      margin: 4px 4px 0 0;
      border-radius: 2px;
    }
  `]
})
export class AppComponent {
  @ViewChild(MousepadComponent) mousepad: MousepadComponent;

  getMouseMessage(m: MouseEvent) {
    return m ? `(${m.clientX}, ${m.clientY})` : '(?, ?)';
  }
}
```

The **async** pipe always returns the last observed value from the data stream. To convert this **MouseEvent** value to an informative text, we use the **getMouseMessage()** function. Because **async** does it for us, we do not need to invoke **subscribe()** on the data stream at all.

Figure 11-10 shows that we always see a single value—the last **MouseEvent** data—when we scribble with the mouse.

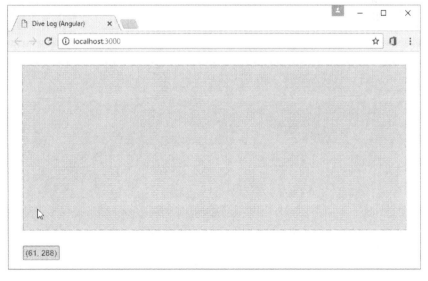

Figure 11-10: The async pipe in action

We can get rid of the **getMouseMessage()** function with a simple template expression trick as the sample in the **Exercise-11-11** folder does:

```
<span class="message">
  ({{ (mousepad.mouseEvents$ | async)?.clientX }},
    {{ (mousepad.mouseEvents$ | async)?.clientY }})
</span>
```

Note, the two **async** operators produce two subscriptions to the **mouseEvent$** stream. If you need to access more properties of an **async**-ed value, apply a conversion function—just like in **Exercise-11-10**—so that you can avoid multiple—and unnecessary—subscriptions to the stream.

Summary

Angular applies the *asynchronous observables pattern*—with its asynchronous data streams—heavily, which pattern is a foundation of *reactive programming*. A data stream is a sequence of ongoing events that are ordered in time. Events can represent three things: a *value*, an *error*, or a *signal of completion*, respectively.

Observables have some useful traits: they are lazy, cancellable, retriable, and composable. To use them, we create an observable stream, subscribe to the stream, and define observer functions to receive data values, error and completion notifications.

Appendix A: Backend Registration Codes

You can utilize the codes in this appendix for using the backend service APIs provided for the exercises in *Chapter 4, REST Operations with the Http Service*.

```
Code 00: 126b8590da81487494320b6d11119daf
Code 01: e1ca948151c149ee87c82a2effe4bf46
Code 02: 8cd5c9a7db2d48ec8c3b4ca9b1cef879
Code 03: b76082d3c24b4608ade3aaa62609a726
Code 04: b4d8fc6f4a2d4d0e9878f46e03820996
Code 05: faabe1ea0ea04db9aeabbabfc2f62cd5
Code 06: 6f629994fef344dbbdcee1a3fe491389
Code 07: 6b0aa0c8a96d433ba948ba6adc75f6f9
Code 08: bcf9e198fb724af6974f566ce8c3d770
Code 09: b9a07219a1db4256bb31dbfdf9e8e103
Code 10: a7870011109a41e7a1f33f75065e60e9
Code 11: 48187130aabf4898992a5c9ae728f0a9
Code 12: f52d1991e3974e07b184a5327b584388
Code 13: d12c5c99de934471a23c285d02d46228
Code 14: 4fa17641d58344869334eb95e2280298
Code 15: b9e920197a0345fe861c1ae758b2a06c
Code 16: e1e6865ea0224ad5ac4069c100a8333c
Code 17: de2d4a4ce4014336b3f7754ef66b66c8
Code 18: 99742acc753c4e2aa8639e39b4ce4166
Code 19: c399aec3347d4b28903904c8d31aa256
Code 20: f9ae709a6e5844e1941934e22eb89367
Code 21: 1c11aa0c77ad43b5adbb3fb7ac9a4984
Code 22: 176039196cd49838a3db6276ebc2e93
Code 23: 5d3db712d34b4a178213fc03f9c89a04
Code 24: 4c4ca90c30464e68aa60e84bffe29f55
Code 25: fa5b841b374047dbba483153b6a5cd44
Code 26: 26502146c5f748d8884e2262729476e4
Code 27: 621d5254eef045099df0a494e85f0b64
Code 28: 6097161809e6405d9bd68205de2bf035
Code 29: 20bea0a08a84432cb371e8524d258d95
Code 30: 5e89bc96f485451e9d31df987575a286
Code 31: 8d568ed7effb4f31a642daaede409915
Code 32: 9e426bcfcae24c3d8d8cce4c52121cdd
Code 33: 1eb878d2e4a94e7e9f4dbde22cd00ab3
Code 34: 0657d8c01d694db080247433fb185373
Code 35: 152a18f04228421bafc9c0339381ec0e
Code 36: 0585e35461fb43c9836ad52717fde51a
Code 37: 1329c22d566044218b99a50c6a9851e4
Code 38: 8fa63380f09543ea917c457deff184e1
```

```
Code 39:  f463bab8f5b14cd9bcdc43aa8f81d465
Code 40:  804b5f0531e8450ea28b85c0ed2bf7a2
Code 41:  ea5097573db64c9ebeb1b81e072befe4
Code 42:  a5c305ffb1e94ff4b4be48d0faa93161
Code 43:  447b3b6a809c498c9c5bb56ad5e1b309
Code 44:  775d3b0e88174a759af3934e6ead7f14
Code 45:  fae099379e3c46a29aa5af1b659fef4b
Code 46:  fc988759bdbb4342bf5b9be724c63655
Code 47:  9abff16d73674f30ac7e311bbbc413dd
Code 48:  dee8f517f5de47ae892bcc555d42d18c
Code 49:  516363396b9547d889faecce169391d3
Code 50:  b7b8faf95b6a4b24a3791fc547954284
Code 51:  70c25a37f8dc491395b5e18a15b6f047
Code 52:  a828d3e486f14a2ba609e2c30812fe1d
Code 53:  6f05973e6efa4d6ba2ab5974484241df
Code 54:  24604af7ad5e43e9a48d6c84c7d3b43d
Code 55:  35f0c0b5141a41d8877a72297ac85bfd
Code 56:  8164d7e119b14550a5b9440b64688c8a
Code 57:  cfed1c1984584338bd37afe6a616eb7a
Code 58:  b2bf0f22585b493db5ec1251f8d40cc8
Code 59:  3468f323bb26475abf09c1ba6f53a1be
Code 60:  47682792f3714e4c8c073d4171d20d00
Code 61:  d84b23fbabb44edcbe53401b6868e544
Code 62:  f429dd7b70314becb9f6a127525a663d
Code 63:  69426638dc1f4cf59aab1ec2bf6f4246
Code 64:  eabf99932d3c4b99a086ae752740bfbe
Code 65:  d18805a781bc4b41bd64f49469631275
Code 66:  0a78b3a928564c508e14069886fec149
Code 67:  ff7f4119bd80422b86218af23d82e3c7
Code 68:  54930b785bae476484449e3266c7c66c
Code 69:  b68b7c1857d04812a1aa300293a3c5ca
Code 70:  dfce1ad5e9684acba188dacc770dd5cc
Code 71:  50ef7a77f27240baab4b88f303aae814
Code 72:  35d3527b38e54821bad61ec6dfd33a70
Code 73:  7fc1562520b7489180bca964d74e0d7f
Code 74:  d2ba907b79a74a8982e791982454d69b
Code 75:  b53c5fb2f8ab4ed79284384b33e4f621
Code 76:  050e8785810a47128fe177b339e03508
Code 77:  adc841107e7d42028fd07be415d06108
Code 78:  d6cf5a3c5f30423d87b60ca453c8686d
Code 79:  e6b3e454075a441fbbaf74b76f2ea3c5
Code 80:  625dde694fb44d63b75a4b84bc650cf7
Code 81:  33c6f83a2f9d4175858641f5e05d1859
Code 82:  5e04d66815cd4e4a9a68e661c5d0b300
Code 83:  9d3cb462548e4165937626528b247889
Code 84:  b26bb89a0316429c876dcc377b85a4d3
Code 85:  1dee0dc9997a489c94e51d4bf37d7f14
Code 86:  aba7f52bf73b47d88ce7606fe0705b18
Code 87:  2eba6112f45340e3b5f620fee90faf8f
Code 88:  4bd0e006f5fb4d0096539f68c8d34acf
Code 89:  707b4dcd3e9e4fb0b6de53e31bc5695a
Code 90:  2578f3acc8e1477a9ef3267e2d2256be
```

```
Code 91: 21be3e3664ea4874af97b78aa52aaad9
Code 92: 448d5c3a1d6f4033b5c2c37103e5b415
Code 93: f6155e2b7f774077947099ad653f067f
Code 94: debb92ee918445a19d685b2fb60d612b
Code 95: 821f96452e7741a6a2dd7f57bb6bdb07
Code 96: 32cb9b5d22b8434fa1f8e0a1938f959e
Code 97: a06672109ldf4d05a3123659998b29ef
Code 98: 0db5dc07d5434c269b8cefb349a2336a
Code 99: f088f39baf2d4dff9e1ce69a0857a28b
```

Appendix B: Preparing and Updating the Components Used in the Book

In this appendix, you will learn how you can update the components used in this book. Here, I explain the way I have installed them. Knowing this may help you update the samples to the latest—or a particular—version of components.

In the samples of the book, I use Angular I tandem with Bootstrap and jQuery. I assume that you have already installed Node.js and its package management tool, **npm**.

I also treat the steps I carried out to create the environment for the samples.

Installing Components

Create a root folder for your samples in your machine, wherever you like. I will refer to this folder as **Samples**. Start a command line prompt, and navigate to the **Samples** folder. Run these commands sequentially after each other in the command prompt:

```
npm init -y

npm install core-js --save
npm install reflect-metadata --save
npm install zone.js@^0.7.2 --save
npm install rxjs@5.0.0-rc.4 --save
npm install @angular/core --save
npm install @angular/common --save
npm install @angular/compiler --save
npm install @angular/platform-browser --save
npm install @angular/platform-browser-dynamic --save
npm install @angular/router --save
npm install @angular/http --save

npm install systemjs --save
npm install bootstrap --save
npm install jquery --save

npm install typescript --save-dev
npm install typings --save-dev
npm install lite-server ---save-dev
npm install concurrently --save-dev
npm install gulp --save-dev
```

```
npm install gulp-util --save-dev
npm install yargs --save-dev
```

Now, you installed all required components. When you look into the **package.json** file, it is like this:

```json
{
  "name": "sample-template",
  "version": "1.0.0",
  "description": "",
  "main": "index.js",
  "scripts": {
    "test": "echo \"Error: no test specified\" && exit 1"
  },
  "keywords": [],
  "author": "",
  "license": "ISC",
  "dependencies": {
    "@angular/common": "~2.3.0",
    "@angular/compiler": "~2.3.0",
    "@angular/core": "~2.3.0",
    "@angular/forms": "~2.3.0",
    "@angular/http": "~2.3.0",
    "@angular/platform-browser": "~2.3.0",
    "@angular/platform-browser-dynamic": "~2.3.0",
    "@angular/router": "~3.3.0",
    "bootstrap": "^3.3.7",
    "systemjs": "0.19.40",
    "core-js": "^2.4.1",
    "jquery": "^3.0.0",
    "reflect-metadata": "^0.1.8",
    "rxjs": "5.0.0-rc.4",
    "zone.js": "^0.7.2"
  },
  "devDependencies": {
    "@types/core-js": "^0.9.34",
    "@types/node": "^6.0.46",
    "concurrently": "^3.1.0",
    "gulp": "^3.9.1",
    "gulp-util": "^3.0.7",
    "lite-server": "^2.2.2",
    "typescript": "^2.0.3",
    "yargs": "^4.7.1"
  }
}
```

Please note, the version numbers of the components may differ from the ones in the listing above.

Modifying the **package.json** file

Change the **scripts** section of the file to this:

```
...
"scripts": {
    "start": "concurrently \"npm run tsc:w\" \"npm run lite\" ",
    "tsc": "tsc",
    "tsc:w": "tsc -w",
    "lite": "lite-server",
    "typings": "typings",
    "postinstall": "typings install"
},
...
```

With these changes, you set up the project to start the TypeScript compiler and **lite-server** concurrently whenever you enter the **npm start** command. Both utilities watch for file changes, so as you change the code, modifications are immediately reflected—after a few seconds—in your default web browser.

Remove the **main** property from the file, and change the package identification properties:

```
...
    "name": "unraveling-angular-2",
    "version": "1.3.0",
    "description": "This project contains the samples of the book",
...
```

With these steps, you have completed the creation of **package.json**.

Creating tsconfig.json

When you run the examples, the TypeScript compiler transpiles the **.ts** files into native JavaScript. Instead of specifying the compiler options in the command line, you describe them in the **tsconfig.json** file. So, create the file in the **Samples** folder to specify these settings:

```
{
    "compilerOptions": {
        "target": "es5",
        "module": "system",
        "moduleResolution": "node",
        "sourceMap": true,
        "emitDecoratorMetadata": true,
        "experimentalDecorators": true,
        "removeComments": false,
        "noImplicitAny": false
    }
```

```
}
```

Creating systemjs.config.js

In order Angular and application modules can be loaded properly, you need to configure the **systemjs** module loader. Create a new file in the samples folder, **systemjs.config.js**, with this content:

```
/**
 * SystemJS configuration file for the Unraveling Angular 2 book
 */
(function (global) {
  System.config({
    paths: {
      // --- Here we define alias
      'npm:': 'node modules/'
    },
    // --- Map Angular modules to their files
    map: {
      // --- All sample apps are within the 'apps' folder
      app: 'app',
      // --- Angular bundels we use in the book
      '@angular/core':
        'npm:@angular/core/bundles/core.umd.js',
      '@angular/common':
        'npm:@angular/common/bundles/common.umd.js',
      '@angular/compiler':
        'npm:@angular/compiler/bundles/compiler.umd.js',
      '@angular/platform-browser':
        'npm:@angular/platform-browser/bundles/platform-browser.umd.js',
      '@angular/platform-browser-dynamic':
        'npm:@angular/platform-browser-dynamic/bundles/platform-browser-
dynamic.umd.js',
      '@angular/http':
        'npm:@angular/http/bundles/http.umd.js',
      '@angular/router':
        'npm:@angular/router/bundles/router.umd.js',
      '@angular/forms':
        'npm:@angular/forms/bundles/forms.umd.js',
      // --- Other libraries we use in the book
      'rxjs': 'npm:rxjs'
    },
    // --- We define how packages should be loaded when no
    // --- filename and/or no extension is defined
    packages: {
      app: {
        main: './main.js',
        defaultExtension: 'js'
      },
```

```
    rxjs: {
      defaultExtension: 'js'
    }
  }
});
})(this);
```

Creating bs-config.json

All samples use components from the **node_modules** folder that is created right within the Samples folder. However, the samples are two level deeper, in the **ChapterNN/Exercise-NN-MM** folder. When you run these samples, lite-server will look the **node_modules** folder within the exercise folder. Add this **bs-config.js** file to the **Samples** folder to resolve this issue:

```
{
    "port": 3000,
    "server": {
        "routes": {
            "/node modules": "../../node modules"
        }
    }
}
```

The **routes** setting tells **lite-server** explicitly to use the **node_modules** folder in **Samples**.

Changing the Configuration Files

As new versions of Angular and its dependent components are released continuously, you may need to change the content of **package.json**, **tsconfig.json**, **bs-config.json** or **systemjs.config.js**. Because this book has more than a hundred exercises, it would be painful and laborious to traverse through all exercise folders and carry out the modifications manually. You can do it easier and quicker:

1. Modify these files directly in the **Samples** folder.

2. Run the **gulp boilerplate** command line from within the **Samples** folder. This command will update the content of each exercise folder with the configuration files you modified in **Samples**.

Made in the USA
Lexington, KY
07 January 2017